Cambridge Elements

Elements of Paleontology
edited by
Colin D. Sumrall
University of Tennessee

A REVIEW AND EVALUATION OF HOMOLOGY HYPOTHESES IN ECHINODERM PALEOBIOLOGY

Colin D. Sumrall
University of Tennessee
Sarah L. Sheffield
University of South Florida
Jennifer E. Bauer
University of Michigan Museum of Paleontology
Jeffrey R. Thompson
University of Southampton and Natural History Museum, London
Johnny A. Waters
Appalachian State University, North Carolina

Shaftesbury Road, Cambridge CB2 8EA, United Kingdom

One Liberty Plaza, 20th Floor, New York, NY 10006, USA

477 Williamstown Road, Port Melbourne, VIC 3207, Australia

314–321, 3rd Floor, Plot 3, Splendor Forum, Jasola District Centre, New Delhi – 110025, India

103 Penang Road, #05–06/07, Visioncrest Commercial, Singapore 238467

Cambridge University Press is part of Cambridge University Press & Assessment, a department of the University of Cambridge.

We share the University's mission to contribute to society through the pursuit of education, learning and research at the highest international levels of excellence.

www.cambridge.org
Information on this title: www.cambridge.org/9781009397179

DOI: 10.1017/9781009397155

First published 2023

A catalogue record for this publication is available from the British Library.

ISBN 978-1-009-39717-9 Paperback
ISSN 2517-780X (online)
ISSN 2517-7796 (print)

A Review and Evaluation of Homology Hypotheses in Echinoderm Paleobiology

Elements of Paleontology

DOI: 10.1017/9781009397155
First published online: March 2023

Colin D. Sumrall
University of Tennessee

Sarah L. Sheffield
University of South Florida

Jennifer E. Bauer
University of Michigan Museum of Paleontology

Jeffrey R. Thompson
University of Southampton and Natural History Museum, London

Johnny A. Waters
Appalachian State University, North Carolina

Author for correspondence: Colin D. Sumrall, csumrall@utk.edu

Abstract: The extraxial–axial theory (EAT) and universal elemental homology (UEH) are often portrayed as mutually exclusive hypotheses of homology within pentaradiate Echinodermata. Extraxial–axial theory describes homology upon the echinoderm bauplan, interpreted through early postmetamorphic growth and growth zones, dividing it into axial regions generally associated with elements of the ambulacral system and extraxial regions that are not. Universal elemental homology describes the detailed construction of the axial skeleton, dividing it into homologous plates and plate series based on symmetry, early growth, and function. These hypotheses are not in conflict; the latter is rooted in refinement of the former. Some interpretive differences arise because many of the morphologies described from eleutherozoan development are difficult to reconcile with Paleozoic forms. Conversely, many elements described for Paleozoic taxa by UEH, such as the peristomial border plates, are absent in eleutherozoans. This Element recommends that these two hypotheses be used together to generate a better understanding of homology across Echinodermata.

Keywords: echinoderm, homology, evolution, anatomy, inheritance

ISBNs: 9781009397179 (PB), 9781009397155 (OC)
ISSNs: 2517-780X (online), 2517-7796 (print)

Contents

1 Introduction

One of the most fundamental factors affecting our ability to understand the evolutionary history of Echinodermata, a long-lived, diverse, and complex phylum of marine organisms, is our ability to identify homologous skeletal elements and regions across different clades. Homology forms the basic underlying assumption set at the root of the morphological data used to infer phylogenetic relationships, allows for understanding the evolution of function, and many other lines of research within the clade. Unfortunately, the homology of skeletal elements is often masked through evolutionary processes that result in character transformations that alter the presentation of characters so that they are unrecognizable, are confused for other morphologically similar character states, or lose their features altogether (Sumrall, 2017). Without an accurate understanding of homology, it is not possible to correctly infer phylogenetic relationships. Identifying homologous elements requires examining morphological features across taxa and through a number of lenses. Extant echinoderms (asteroids, ophiuroids, holothuroids, echinoids, and crinoids) evolved early in the group's history and offer insights from larval morphologies, development, and gene expression, but are strikingly different from the extinct clades of Paleozoic echinoderms, such as blastozoans and homalozoans that have no modern representatives. Paleozoic forms have greater taxonomic and morphologic diversity, but the high levels of convergence, and disparate bauplans throughout their evolutionary history (Ubaghs, 1971; Sumrall, 2017; Deline et al., 2020) have complicated our ability to unravel the origins, ontogeny, evolution, and life mode of these animals.

Echinoderms have a biomineralized skeleton of mesodermal origin (see Gorzelak, 2021). The development of the biomineralized echinoderm skeleton is associated with a distinct set of transcription factors, signaling molecules, and differentiation genes, which all act together during development and comprise the biomineralization toolkit of echinoderms (see review in Thompson 2022). It is the expression of the components of this biomineralization toolkit in particular cells, typically referred to as skeletal cells, which underlies skeletal growth and development, and many skeletal cells occupy the porous cavities that characterize echinoderm stereom (Czarkwiani et al., 2016; Thompson, 2021). The location and molecular fingerprint of these skeletal cells is controlled by distinct spatial signaling cues which are sent from the ectoderm (Duloquin et al., 2007, Czarkwiani et al., 2021). It may be that the activity of these signaling molecules in distinct tissues may result in the presence or absence of skeleton in particular portions of the body wall (Zamora et al., 2022). However, while it has been demonstrated that different components of the echinoderm biomineralization toolkit are expressed in different biomineralized structures (Piovani et al., 2021), there is so far no evidence to indicate that the expression of distinct biomineralization genes is associated with particular skeletal plates that could form the basis for a homology scheme.

Patterson (1982) proposed three tests to falsify hypotheses of homology: two a priori tests (similarity and conjunction) and one a posteriori test (congruence). The test of similarity proposes that hypotheses of homology between two structures could be supported if they are similar in construction. The test of conjunction proposes that hypotheses of homology between two structures assumed to be a singular character transformation are falsified if those two structures are both present in the same organism. The a posteriori test of congruence states that if a character transformation appears more than once optimized on a phylogenetic tree, the feature must have evolved more than once and is therefore rejected as homologous. For a full discussion of echinoderm homology examples of each of these tests, refer to Sumrall (1997).

Difficulties in identifying homologous elements are further compounded by human efforts to define evolutionary relationships. Long-standing methods of delineating taxonomic groups, dating back to the first attempts at classification (Linnaeus, 1758), emphasized differences among taxa rather than emphasizing similarities that can be used as evidence to recognize taxa. Further, taxonomists have defined distinct and often conflicting sets of terminology for individual groups, making it difficult to discuss homologous elements that groups might share (Sumrall, 2017). This concept is prominent in Paleozoic echinoderm clades, where many classes have different and incompatible lexicons of morphological terms describing homologous morphology. Horizontal comparisons of terms show these include multiple names for homologous morphologies and the same term being used for a variety of nonhomologous morphologies (Sumrall, 2010; Sumrall and Waters, 2012; Sheffield and Sumrall, 2019; Ausich et al., 2020).

Additionally, there are issues with definition and diagnosis, similar to Rowe's (1988) discussion of these terms with respect to a clade, when applying many homologous terms to blastozoans. Rowe (1988) used definition to describe a clade as a historical entity based on ancestry, for example, a most recent common ancestor and all descendants. This is distinct from a diagnosis, which is used to identify group membership based on attributes. In practical terms, for example, glyptocystitoid rhombiferans are often diagnosed by a series of features such as the plating of the theca, the presence of dichoporite respiratory structures, and morphologies of the stem (Zamora et al., 2017). Because such diagnostic structures are not universally found among all taxa, including stem lineages arising prior to their evolution and character losses in derived taxa, a simple diagnosis is not sufficient to circumscribe all relevant taxa. This creates diagnoses such as "feature present unless secondarily lost," which requires a priori knowledge of the phylogenetic placement of a taxon to diagnose it as a member of the clade. However, by defining taxa based on shared ancestry within the context of an evolutionary tree, we can circumscribe them as a clade and the presence of diagnostic traits becomes irrelevant to our understanding of their group membership (Brochu and Sumrall, 2001). Optimized onto the phylogeny, diagnostic traits can be seen to evolve within a series of nested clades and these traits are evidence used to understand the structure of the tree. But, modern phylogenetic methods define taxa based on tree structures, rather than as objects that bear suites of characters.

In many respects, homology can be seen in a similar framework. Because homologous structures can be defined by their evolutionary origin as a synapomorphy and have a fate as character state transformations in descendant lineages, clades are recognizable by bearing diagnosable alternate states in the character description. Homology is defined based on the historical origin of structures – if structures are derived from a common ancestral origination, such as a bird wing and a whale flipper are, then at the level of forelimbs they are homologous. In practice, however, we often rely on diagnosis, where we assume homology because two structures have a common construct that can be identified. We can list the features that we expect a given structure to have and, if they are present, then we assume the homology, but true homology can only be understood within a historical framework.

Given the lack of developmental information in many cases, inferring homology based on a diagnosis is indeed problematic, but often unavoidable. Homology is clearest where there is ontogenetic information, and tracing specific skeletal elements throughout ontogenetic stages provides strong evidence for their homology and their identification in mature specimens. The plates that cover the mouth in most echinoderms, the primary peristomial cover plates (PPCP), are present in all identifiable early postlarval taxa, where known, including edrioasteroids, crinoids, blastoids, and a host of other stemmed echinoderms (Sumrall and Waters, 2012). These plates can be traced ontogenetically and are often quite evident in mature specimens of these taxa. This information can then be applied to other taxa that descend from a common ancestor. In other cases, such as plating of the axial skeleton, the developmental pathway can be seen in the presence of terminal growth of the axial plate series. Of course, there are clear limitations concerning ontogenetic data in the fossil record, as many groups of fossils are not represented by different growth stages (e.g., many diploporans have few documented juvenile forms; Sheffield et al., 2018).

In EAT (see below), the imperforate and the perforate extraxial skeleton are treated as separate entities, defined in recent taxa by their origination, growth, and development, but in fossils they are diagnosed by their character expression, which is subject to heterochrony and heterotopy. Consequently, inferring homology based on a diagnosis is not universally reliable. Numerous other examples in Paleozoic echinoderms indicate that the distinctions between perforate and imperforate extraxial skeleton may have little to do with the underlying developmental pathways but are diagnosed instead by the expression of ephemeral morphological features (Fig. 1). In what follows, we analyze homology hypotheses for major features of echinoderm bodies: (1) respiratory structures; (2) feeding structures; (3) hydropores, gonopores, and periprocts; and (4) oral surface plating. As UEH (see below) was developed for plates of the oral area and ambulacral system, we cannot provide an analysis of each of these major features through both UEH and EAT.

This Element, focusing on homology hypotheses, requires a grasp of echinoderm morphologies. It is outside the scope of this review to introduce the details of morphology and body plans for the major echinoderm groups we discuss herein. Here, we provide references focusing on morphological features and body plans for these major groups. As echinoderm morphology is highly disparate, we refer readers to a large body of literature: Blastozoa (Sprinkle, 1973; Sumrall and Waters, 2012; Sheffield et al., 2022); Crinozoa (Kammer et al., 2013; Ausich et al., 2020); Echinozoa (Smith, 1984a; Kerr and Kim, 2001); and basal echinoderms (Parsley, 1980; David et al., 2000; Smith 2005; Zamora et al., 2012; Zamora and Rahman, 2014).

There are two foundational hypotheses for understanding echinoderm homology: (1) extraxial–axial theory (EAT; Mooi et al., 1994; Mooi and David, 1997; David and Mooi, 1998) and (2) universal elemental homology (UEH; Sumrall, 2010; Sumrall and Waters, 2012). The EAT hypothesis is built upon parameters of growth and development in extant echinoids and provides homology designations for different skeletal regions of the echinoderm body plan. This coarsely divides the echinoderm body plan into two regions differentiated upon their mode of growth: the axial region (essentially, the ambulacral system and related structures) and the extraxial skeleton (Fig. 2). The extraxial skeleton is further subdivided into the perforate and the imperforate extraxial skeleton based upon the presence or absence of piercings of the body wall. From the fossil perspective, the UEH hypothesis is built to recognize homology of individual skeletal elements of the peristomial border and axial skeleton across different groups of echinoderms. In essence, the UEH hypothesis refines homology

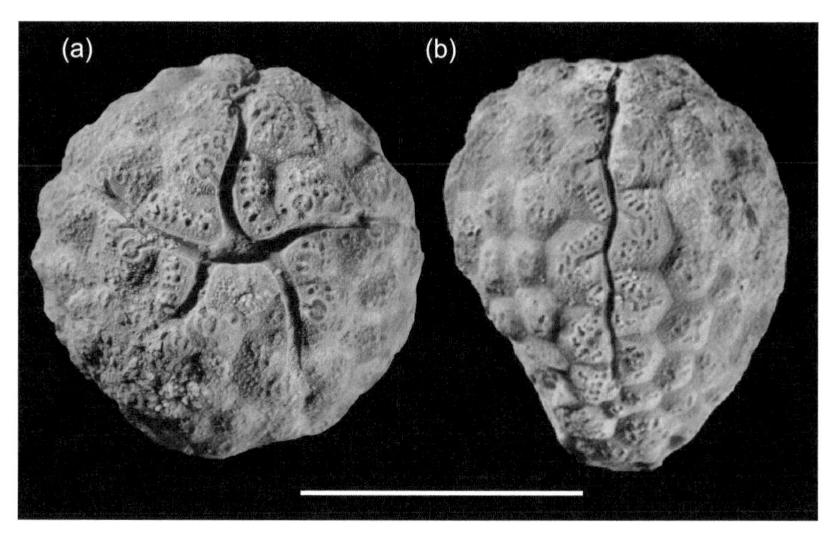

Figure 1 Respiratory structures piercing axial skeleton. (a) Oral view of *Estonocystis antropoffi* (GIT 540–80). The five ambulacral grooves lie on oral plate sutures; brachiole facets begin after the oral plate series and continue down the theca. (b) Side view of *Estonocystis antropoffi* (GIT 540–80). The ambulacra anastomose down the theca; short food grooves that connect with the main food groove lead to brachiole facets. These brachiole facets are contained within the center of single ambulacral floor plates, and diplopores align horizontally along the floor plates. In EAT, diplopores should be contained within perforate extraxial plates of the theca, while floor plates belong in the axial system. Both are modified from Sheffield and Sumrall (2019). Specimen whitened with ammonium chloride sublimated. Scale bar = 10 mm.

of the axial skeleton so that commonalities among plates of the oral region and ambulacra can be understood across pentaradial echinoderms. This system is limited by the lack of recognizable homologous features in the oral area of fossil eleutherozoans and homalozoans and scant knowledge of the earliest stages of development from extinct taxa. Extraxial–axial theory and UEH are often discussed as frameworks that exist in opposition to one another; we clarify here that they are compatible and capable of complementing one another to better our understanding of echinoderm homology (Sumrall and Waters, 2012). Herein, we provide a thorough review of both EAT and UEH and offer new insight on combining the two approaches. We also review future research directions utilizing these homology hypotheses with echinoderm taxa bearing unusual morphologies.

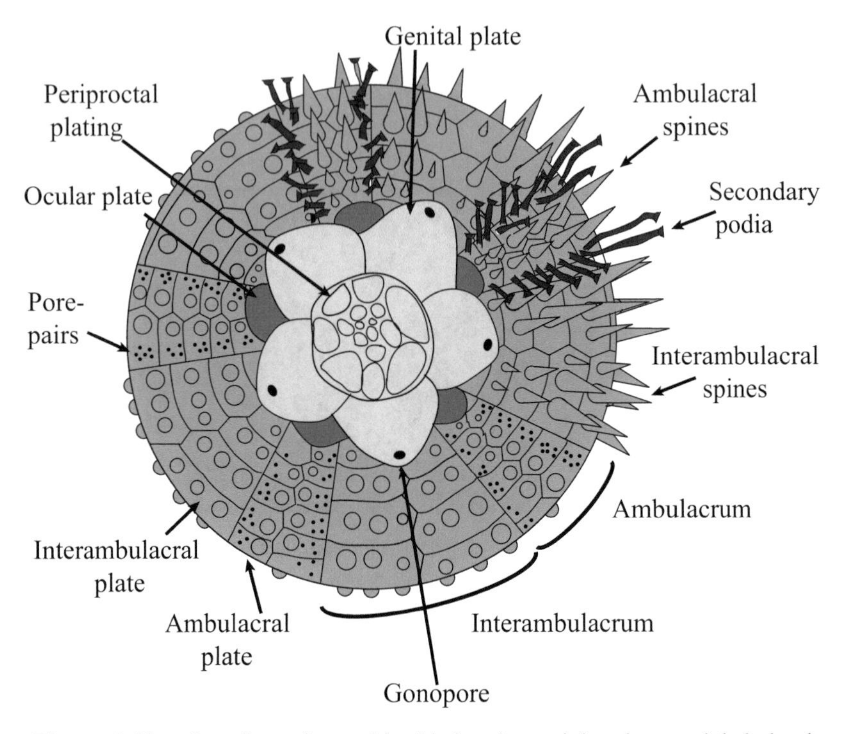

Figure 2 Aboral surface of an echinoid showing axial and extraxial skeletal elements as delimited by the EAT. Axial tissues are shown in shades of gray, while extraxial tissues are in blue. Morphological structures of note are highlighted with arrows. Modified from Savriama et al. (2015) and Thompson et al. (2021).

1.1 Institutional Abbreviations

British Museum of Natural History, London (BMNH), now the Natural History Museum, London (NHMUK); Cincinnati Museum Center (CMC IP); Geological Institute of Tallinn (GIT); Museum of Paleontology, Guizhou University, Guiyang, China (GM); Prague National Museum (L); Museo Geominero (Geological Survey of Spain), Madrid, Spain (MGM); Paleontological Institute of the Russian Academy of Sciences (PIN); University of Iowa (SUI); Texas Memorial Museum (TX); and University of Michigan Museum of Paleontology (UMMP).

2 Homology Hypotheses for Echinodermata

Homology, as it relates to fossil organisms, is often explored from a comparative anatomy framework to determine similarity or adaptive sequences of structures (Patterson, 1982; Wagner, 2007; Wright, 2015). The framework of developmental genetics, which is actively investigated among modern echinoderms (Oliveri et al., 2008; Shashikant et al., 2018; Thompson et al., 2021; Thompson, 2022), further complicates such studies in the Paleozoic, as many of the organisms are in extinct clades and different or similar genetic pathways may have produced nonhomologous structures that appear homologous (Shubin and Marshall, 2000; Shubin et al., 2009; Wright, 2015). Ground truthing our a priori interpretations of homologous structures among extinct forms requires the application of a posteriori phylogenetic methods.

One of the two major homology hypotheses, extraxial–axial theory (EAT), differentiates the echinoderm skeleton into two broad categories: the axial (associated with the oral and ambulacral system, including morphologies covered by UEH) and the extraxial (the nonambulacral body wall, subdivided into perforate and imperforate regions; Mooi et al., 1994; Mooi and David, 1997, 1998, 2008; David et al., 2000). The second hypothesis, universal elemental homology (UEH) (Sumrall, 2010, 2017; Sumrall and Waters, 2012; Ausich and Kammer, 2013; Kammer et al., 2013), is a framework to identify the homologies of the individual skeletal elements of the oral area and ambulacral system within echinoderms. Although some authors have tended to treat EAT and UEH as alternate hypotheses for homology among echinoderms (e.g., Guensburg and Sprinkle, 2001, 2007; Zamora and Rahman, 2014; Sheffield and Sumrall, 2019), or to favor one hypothesis without considering the other (Paul, 2021), these hypotheses address different aspects of homology and can be used in concert. Each hypothesis articulates observations about homology from different aspects of the echinoderm bauplan. Extraxial–axial theory is regional, covering the entire organism. It was applied among echinoids and other Eleutherozoa and later applied and interpreted across Pan-Echinodermata (sensu Sumrall, 2020). Universal elemental homology is a high-precision hypothesis, but it is limited to the mouth frame and ambulacral system. It was applied among blastozoans (including crinoids; Ausich and Kammer, 2013; Kammer et al., 2013; O'Malley et al., 2016) and edrioasteroids; because of a lack of morphological framework across disparate groups, UEH has not been expanded to eleutherozoans and homalozoans. These two hypotheses, EAT and UEH, are not in conflict, nor are they alternatives to one another (Smith and Zamora, 2013; Wright, 2015). Instead, they complement one another when considering a holistic approach to echinoderm homology (e.g., body regions and specific skeletal elements).

2.1 Extraxial–Axial Theory

Extraxial–axial theory hypothesizes skeletal homologies across broad regions of the echinoderm body plan and characterizes them based on their presumed developmental origin. It divides the plating of the body wall into axial and extraxial elements. The axial region of the skeleton is associated with the water vascular system, as it radiates from the peristome and grows from the distal tip, adding new elements by terminal addition. The perforate extraxial region of the skeleton bears elements that can be inserted at multiple points without adherence to terminal addition. It contains several body openings including the periproct, hydropore, gonopore, and a variety of respiratory structures. This strict diagnosis interprets the oral plates and oral frame plates as being part of the perforate extraxial skeleton, but this is inconsistent with their development and association of these elements with the floor plate system, and the fact that the body openings associated with them occur only in more derived taxa. It is also important to note that in *Kailidiscus*, oral plates and the precursor plates to the oral frame elements bear podial pores that are only found in axial skeleton (Zhao et al., 2010). If one infers their homology strictly on diagnosis, they can be either axial or extraxial, but phylogenetically and developmentally they are axial.

The imperforate extraxial skeleton lacks the pore systems associated with the perforate extraxial skeleton (Mooi et al., 1994). These regions were identified based upon the different modes of growth in different aspects of the echinoid skeleton, then associated with the divergent larval origin of different aspects of the adult body plan and then translated to other echinoderm clades (Mooi and David, 1997; David et al., 2000). In practice, skeletal type is diagnosed by features of the growth parameters and pore types, which are subject to heterochrony (absence of pores in paedomorphic taxa) and heterotopy (evolutionarily relocating structures). Consequently, this diagnosis-based system tends to rely heavily on features of these plate fields (ocular plate rule (OPR), presence of pores) in extinct clades where developmental evidence is more difficult to interpret. This is why the floor plate series of diploporans like *Dactylocystis* and *Tristomiacystis*, which grow by terminal addition and are developmentally composed of ambulacral floor plates but also bear respiratory structures, are difficult to reconcile. One must choose whether to diagnose them as axial, based on OPR, or perforate extraxial, based on pore systems; they cannot be both. Homology would require us to accept the developmental argument over the diagnosis argument and infer them to be ambulacral floor plates.

Axial skeleton is positioned in the rays and is generally associated with the water vascular system. Axial elements are recognized by the OPR, in which plates are added to the growing tip of the plate series immediately proximal to a terminal ossicle or ocular plate. In practice, this precise diagnosis does not work for most echinoderm taxa, as ocular plates are documented with certainty only in crown group Eleutherozoa and are demonstrably absent in nearly every other echinoderm clade. Some authors have described the OPR in noneleutherozoans (e.g., Smith, 1985; Nohejlová et al., 2019; Paul and Toom, 2021); however, these interpretations have not been convincing to the authors as definitive evidence for the OPR. Instead, terminal growth is implicitly used to identify the axial skeleton presumably as an expression of the ocular plate rule. But issues exist with this simple distinction. Early crinoid arms are compound structures that have axial floor plates abutting brachial elements that are classified as extraxial skeleton, both of which grow by terminal addition (Guensburg et al., 2015). In more derived crinoids, extraxial brachial elements commonly become biserial, mimicking the ocular plate rule, and otherwise are indistinguishable from true axial skeleton outside a phylogenetic framework (Kammer et al., 2013). In some taxa of Pennsylvanian cladid crinoids, such as *Erisocrinus*, the brachial elements begin as uniserial in juvenile ontogenetic stages and then transition to biserial wedges in older stages (Peters and Lane, 1990; Sheffield, 2013). Furthermore, within eleutherozoans, the nonambulacral marginal frame of asteroids and the extinct somasteroids are added via terminal addition next to the ocular plate, casting doubt on whether or not these structures are axial or extraxial (Mooi and David, 2000; Hotchkiss, 2012).

Many more complex plating arrangements exist in the axial skeleton, including at least three different plate series: adradial floor plates, abradial floor plates, and cover plates (Fig. 3) (Zhou et al., 2010; Sumrall, 2015, 2017). Within these plate series, axial skeleton does not always follow simple terminal growth. In derived blastozoans, such as blastoids, glyptocystitoids, and hemicosmitoids, the abradial floor plates are complex and are differentiated into a primary and a secondary series (Sprinkle, 1973: fig. 4; Sumrall, 1997).

Not all axial skeleton follows the ocular plate rule. Other than those that form a simple biseries, cover plates do not always develop by terminal addition. In edrioasteroids, distal cover plates are generally arranged into a simple biseries, but with maturity, secondary and even tertiary cover plates are added later in ontogeny in more proximal regions of the ambulacral system (Figs. 4, 5) (Bell, 1976b; Bell and Petersen, 1976). It is possible, however, that these plates are small and internal and only expressed later in ontogeny (for a discussion, see Sumrall, 1996: p. 970). The remainder of the skeleton that is not defined as axial is considered extraxial (Mooi and David, 1997, 1998, 2008; David et al., 2000).

The extraxial skeleton roughly equates to the interambulacral plating of nonechinoids: thecal wall plating, stem, and holdfast. Such plating is generally irregular in early taxa such as edrioasteroids and eocrinoids (Fig. 4), as opposed to the more derived and highly organized theca of later echinoderms such as blastoids, crinoids, and glyptocystitoids. Perforate extraxial skeleton is generally more proximally positioned and defined based on the presence of pores or

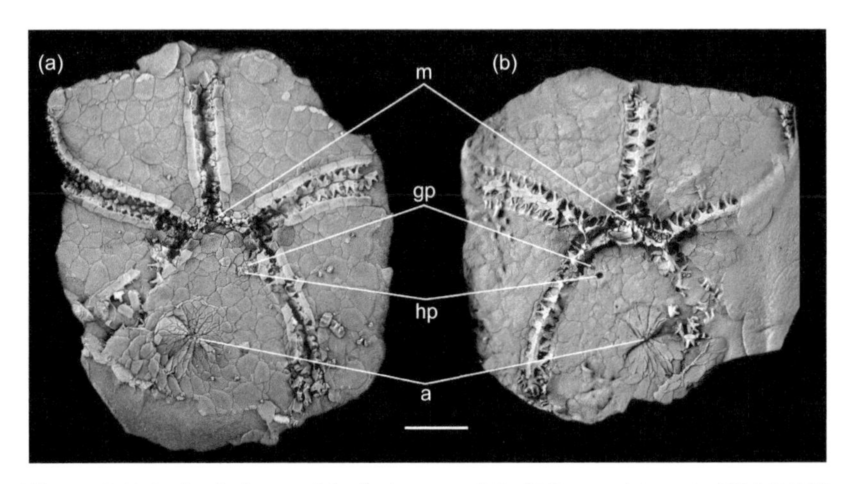

Figure 3 Colorized views of the holotype of *Kailidiscus chinensis* (GM 3428) showing the morphology of the axial skeleton. (a) Exterior view with most of the cover plates stripped. (b) Interior view. Teal = abradial floor plates; green = adradial floor plates; red = oral plates; purple = precursor to oral frame plates; yellow = ambulacral cover plates. a = anus; gp = gonopore; hp = hydropore; m = mouth. Note the position of the ambulacral pores and how the oral plates are contiguous with the abradial floor plates and the oral frame plate precursors are contiguous with the adradial floor plates. Modified from Zhao et al. (2010). Scale bar = 5 mm.

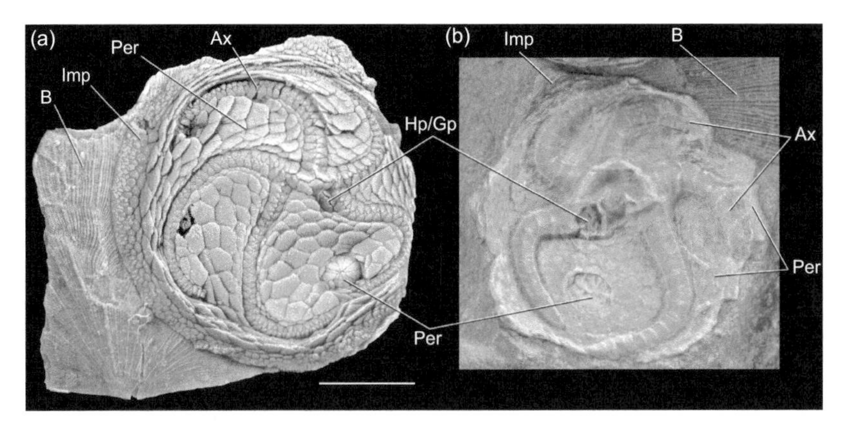

Figure 4 (a) Exterior and (b) interior views of the isorophid edrioasteroid *Isorophus cincinnatiensis* (CMC IP 34539 and CMC IP 23536, respectively) showing the distribution of skeletal types. The ambulacral system, including cover plates, oral frame plates, and orals, are axial skeleton (Ax) = green, the interambulacral plating is perforate extraxial skeleton (Per) = yellow, the peripheral rim is imperforate extraxial skeleton (Imp) = orange, the hydropore/gonopore (HP/GP) = purple, and periproct = blue; they perforate the interambulacral plating. Scale bar = 5 mm.

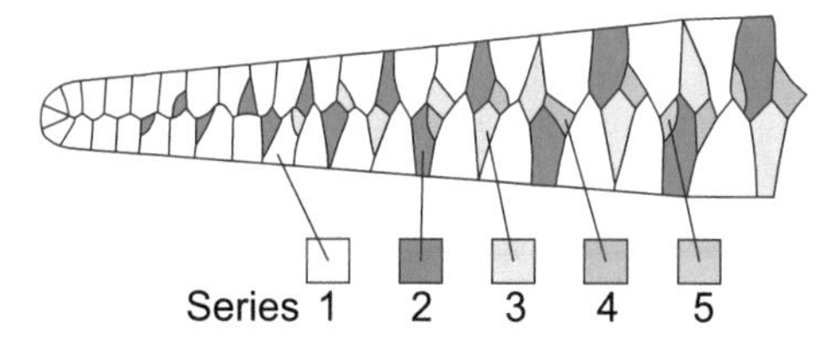

Figure 5 Ambulacral cover plate ontogeny of the edrioasteroid *Postibulla lukei* showing the insertion of plates along the arm rather than purely terminal growth (redrawn from Bell and Petersen, 1976: figure 6). Note that on the distalmost tip of the ambulacral cover plate series (left) the primary cover plate series (white) are added by terminal growth. Higher-order cover plate series (coded by labeled colors) are inserted between plates of lower order along the perradial suture.

thecal openings. In echinoids, this has been equated to the genital and periproctal plating, and the oral and oral frame plating. Imperforate extraxial skeleton is generally more distally positioned and lacks these pores (Mooi et al., 1994).

Much has been made of the supposed molecular and developmental basis for the EAT. David and Mooi (1998) proposed that the extraxial and axial elements of the body wall have different developmental origins resulting from metamorphosis. They proposed that the axial skeleton is associated with tissue arising from the larval rudiment, while the extraxial skeleton is associated with those portions of the adult body plan which are derived from nonrudiment tissue of the larvae. This was followed up with a later comparison of the spatial relationship of the axial and extraxial tissue to the expression of a few homeodomain-bearing transcription factors (Mooi et al., 2005). These gene expression patterns used to support the EAT, published by Lowe and Wray (1997) and discussed by Mooi et al. (2005), are largely well known to be transcription factors associated with the nervous system (as pointed out by Mooi et al., 2005), and thus their expression in a pentaradial pattern in association with the axial skeleton is more likely a function of their expression in the development of the nervous system rather than the development of the skeleton. Further attempts were made to link the extraxial and axial regions of the skeleton to the translocation and expression during development of *Hox* genes (Mooi et al., 2008; David and Mooi, 2014). Any relationship between the expression or translocation of *Hox* genes and the patterning of the adult pentaradial body plan has, however, also been widely refuted (Byrne et al., 2016). At present, there remains little in the way of gene expression patterns that supports any particular homology scheme, EAT or UEH, among echinoderm groups.

The power of EAT lies in its ability to identify areas of regional homology. This facilitates gross morphological characterization of the theca and its

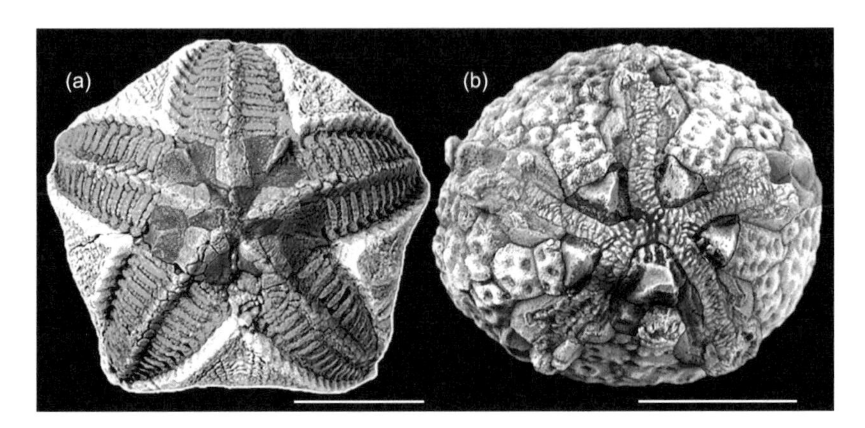

Figure 6 Blastozoan homologies through a UEH framework. Each color corresponds to a specific plate type (false colorization), hypothesized to be homologous, even if incorporated into the body in different manners. Red = oral plates; blue = primary peristomial cover plates; tan = ambulacral cover plates; green = ambulacral floor plates; yellow = thecal plates. (a) Parablastoid *Eurekablastus ninemilensis* (1781TX5; modified from Sumrall 2017). (b) *Eumorphocystis multiporata* (SUI 97598; modified from Kammer et al., 2013). Scale bars = 5 mm.

structures into a framework of homologous regions, and changes in their distribution and characterization can effectively be used to generate phylogenetic characters for analysis. However, its framework lacks the precision required for detailed morphological descriptions of plate arrangements within the ambulacral system and thecal wall. This limitation results in simplified morphological characters, such as counts of the number of basal plates, radial plates, and oral plates, which must be used with extreme caution. In the past these characters have been used in phylogenetic analyses without regard to whether the plates in question were homologous and how they related to the animal's body axes (for examples, see Smith, 1984b; Sumrall, 1997; Frest et al., 2011). This produces characters based on gross similarity rather than homology, and this shortcoming directly led to the development of the UEH hypothesis for plate homologies.

2.2 Universal Elemental Homology

Universal elemental homology was originally developed to address plate homology problems in blastozoans by examining the growth and development of the peristome and ambulacral system and determining the exact identity of skeletal elements across taxa. Subsequently, it was applied to these taxa, avoiding the nonrecognition of homologous structures that plagues the expansion of EAT homologies from echinoids into blastozoans. The application of UEH begins with identifying the plesiomorphic symmetry of the pentaradial echinoderm – what Sprinkle (1973) termed the 2–1–2 symmetry (also see Sumrall, 2010, 2017; Sumrall and Waters, 2012; Kammer et al., 2013). In the plesiomorphic state, three ambulacra exit the peristome: the anterior A ambulacrum and the lateral shared ambulacra, BC to the right and DE to the left. Bifurcation of these shared ambulacra form the distal B, C, D, and E ambulacra (Sumrall and Wray, 2007). Two different series of plates can form the border of the peristome: the oral frame plates, which are plesiomorphically radially positioned, internally expressed, and form the proximal-most plate in the adradial floor plate system. Oral plates are plesiomorphically interradially positioned, broadly expressed externally, and form the proximal-most plate in the abradial floor plate series (Fig. 6) (Sumrall, 2017). Patterson's (1982) test of conjunction confirms these two plate series cannot be homologous, as there are examples of a few taxa that have both oral plates and oral frame plates (Kammer et al., 2013).

Distal to the oral area are floor plates that form the food groove in most taxa. Two types of floor plates with different expressions are found in the axial skeleton and correspond with the types of peristomial bordering plates present (Sumrall, 2017). These two plate series are present in several taxa, documenting that they are not homologous by the test of conjunction (Patterson, 1982; Zhao et al., 2010; Sumrall and Zamora, 2011; Sumrall and Zamora, 2018). Adradial floor plates are typically internally expressed and are dominant in taxa that bear oral frame plates. Abradial floor plates are broadly expressed externally and are dominant in taxa bearing oral plates (see Zamora and Sumrall, in press). Indeed, the oral plates and oral frame plates appear to be the proximal-most plates in these respective floor plate series on morphological grounds.

The peristomial opening is covered by five primary peristomial cover plates (PPCP) that form early in ontogeny (Sumrall and Waters, 2012; Kammer et al., 2013) and are positioned interradially. The PPCPs can be traced ontogenetically (Bell, 1976b; Sumrall and Wray, 2007), and in some taxa they remain prominent after they reach maturity and are easily distinguishable from the surrounding ambulacral cover plates; this feature can be seen in coronoids. The primary peristomial cover plates can become indistinguishable from the shared cover plate and proximal-most ambulacral cover plate systems except by position as in most isorophinid edrioasteroids. Shared cover plates are often present over the peristome, and ambulacral cover plates extend down the floor plate system and protect the food groove.

The power of UEH lies in its ability to identify the evolutionary fate of the development of individual skeletal plates within plate series of the axial skeleton. For example, the loss of one or more ambulacra has evolved in different clades, such as glyptocystitoids, hemicosmitoids, and paracrinoids (Sumrall and Wray, 2007). By understanding the identity of plates present in the oral area and which ambulacral bound them, it is possible to determine the evolutionary fate of each of the ambulacra and code for presence or absence of homologous elements, rather than simply counting the appendages, which are subject to homoplasy. This detailed understanding of ambulacral identity is only possible when the plating of the oral area is fully characterized through UEH, because other orientation features such as positioning of the hydropore, gonopore, and periproct in the CD interray are not fully consistent among echinoderms (Sumrall and Wray, 2007; Sumrall, 2010; Sumrall and Waters, 2012).

Universal elemental homology is useful in providing a more comprehensive understanding of individual homologous elements related to the oral and ambulacral plating in blastozoan echinoderms but has not been expanded to other sections of the body, such as respiratory structures or the thecal body wall plating (extraxial skeleton). Additionally, UEH has not been successfully applied to nonblastozoan or edrioasteroid echinoderms to date.

3 The Application of These Homology Hypotheses

There are several limitations when applying EAT and UEH to blastozoan echinoderms. Because EAT was developed using eleutherozoans as exemplars, where developmental information is more available, there has been an imprecise translation of skeletal regions to noneleutherozoan taxa. Similarly, because UEH was first developed to describe blastozoan mouth frames, this hypothesis can be difficult to reconcile with eleutherozoans, where mouth frame construction is radically different. These limitations reflect the high disparity between these taxa and are evident in the complexities in coding blastozoan character data for eleutherozoans, where a large proportion of the states are mutually inapplicable (Deline, 2021).

3.1 Perforate Skeleton

While a diagnostic feature of perforate extraxial skeleton is the presence of perforations through the thecal wall, perforations are not universally present; they change through ontogeny, and there are issues associated with what is meant by perforations across echinoderms, as described previously. The nature of interambulacral plating in edrioasteroids is a clear example of this complication where the extraxial plating between the axial elements of the ambulacra shows a wide variety of expressions (Bell, 1976a; Zhao et al., 2010). They are inferred to be homologous a priori based on similarity arguments as well as a posteriori based on phylogenetic arguments (Smith and Jell, 1990, following Patterson, 1982). Furthermore, the development of this plate series is well constrained in both early taxa (Zhao et al., 2010) and later taxa (Bell, 1976b; Sumrall, 2001), and there is little doubt that this plate series is homologous across the clade and likely beyond this clade, such as the oral surface plating in imbricate and gogiid eocrinoids. The issue with imperforate and perforate extraxial skeleton is the way they are diagnosed in the absence of direct developmental data. Homology is when two structures are the same historically because they can be traced back to a single structure in a common ancestor, whereas perforate and imperforate skeleton are diagnosed by their character expression. But it is also problematic that a variety of nonhomologous structures are considered as evidence for perforation including the hydropore, gonopore, and periproct (important body openings) – as well as a variety of nonhomologous respiratory structures that either perforate (typically exothecal respiratory structures such as epispires, though there are exceptions) or invaginate rather than perforate the thecal wall (typical of endothecal respiratory structures). These are in addition to pores associated with the podia. Presence of these features, primarily the respiratory structures, is controlled by heterochrony, and the position can vary considerably because of heterotopy. In the context of the development of these structures, it is not inconceivable that signaling from the developing soft tissue structure (be that podia, epispires, etc.) to the growing skeleton is responsible for the presence of perforations in numerous nonhomologous skeletal structures.

3.1.1 Respiratory Structures

Homology becomes most complex where plate series or thecal regions are diagnosed strictly on morphological grounds without respect to underlying developmental criteria. Perforate extraxial skeleton is diagnosed by two factors: (1) not following the OPR (terminal growth); and (2) having perforations in the integument (Mooi et al., 1994). Having perforations in the integument is a complex issue and is a function of several competing factors. First, not all perforations of the integument are homologous nor are they reflecting the same organ systems of the body. Second, some organ systems, such as respiratory pores, change their morphological expression ontogenetically. This suggests that pore systems can be strongly influenced by heterochrony, such as descendent lineages that have pores giving rise to paedomorphic descendants that lose those pore structures. As functions such as respiration are dependent on surface area, which ontogenetically increases more slowly than volume, we see countless examples of evolutionary adaptations of organisms increasing efficiency of respiration (McKinney and Sumrall, 2011). An example of this is seen in the blastoid *Pentremites*, where there is documented evidence that hydrospire respiratory structures grew with positive allometry in order for the surface area of the respiratory structures to keep pace with the volume (Dexter et al., 2009). This has also been documented in the rhombiferan *Pleurocystites*, whose pectinirhombs also grew with positive allometry (Brower, 1999). These variations can range from heavier respiratory structure concentrations in certain areas of the body to maximize efficiency, losing respiratory structures altogether, or developing them at different ontogenetic stages.

It has been well documented that the array of different types of respiratory structures cannot be homologous. First, their construction is vastly different in the groups in which they are present (Patterson, 1980: test of similarity). This includes endothecal respiratory structures, where ambient water is passed through canals embedded within the skeleton for gas exchange through thin stereom folds, and exothecal respiratory structures, where coelomic fluid circulates through the skeleton toward the theca surface (in some cases making true skeletal perforations) for gas exchange (Sumrall and Waters, 2012). Phylogenetically, these are the derived condition in many clades (Patterson, 1980: test of congruence), suggesting that these features cannot be homologous. In fact, some groups such as glyptocystitoids have several different types of respiratory structures within the clade, including both endothecal and exothecal types (Paul, 1968a,b; Sprinkle and Wahlman, 1994; Zamora et al., 2017), and other taxa, such as eublastoid *Troosticrinus*, possess both endothecal and exothecal respiratory structures in the same organism (Sumrall and Waters, 2012). For a comprehensive review of respiratory structures in many Paleozoic echinoderms, we refer readers to Sheffield et al. (2022).

Groups of early edrioasteroids, such as cambrasterids and stromatocystitids, have epispires along the plate sutures in this plate series (Zamora et al., 2007; Zhao et al., 2010). These epispires may be absent in juveniles and become more pronounced ontogenetically. Other taxa, such as the early form *Kailidiscus* and edrioasterids, and later isorophids, lack these structures and bear imperforate interambulacral plating. Yet, within isorophids, the unusual *Thresherodiscus* reevolved respiratory structures in the form of paired pores within plates that are connected by a thin, calcified, hollow bulb similar to those of some kinds of diplopores (Sumrall and Gahn, 2006). The ephemeral nature of respiratory structures in interambulacral plating shows that the situation is more complicated than a simple dichotomy of pores being present or absent.

The addition of respiratory structures during ontogeny has been documented in gogiid eocrinoids, such as *Sineocrinus*, *Guizhouecrinus*, *Akadocrinus*, glyptocystitoids, and other taxa (Fig. 7) (Sumrall and Schumacher, 2002; Nohejlová and Fatka, 2016; Sheffield et al., 2022). In gogiids, epispires are completely absent from the most juvenile specimens, typically those with thecae under 3 mm in height (Fig. 7(a)) (Parsley and Zhao, 2006). It is likely that the juveniles respired across the plates or through the gut, until larger size necessitated epispire development. Because small juveniles have extremely high surface-area-to-volume ratios, it is not unexpected that juvenile blastozoans lacked respiratory structures and instead were able to respire across the plates, whereas this ratio dramatically decreases with increased size, requiring respiratory

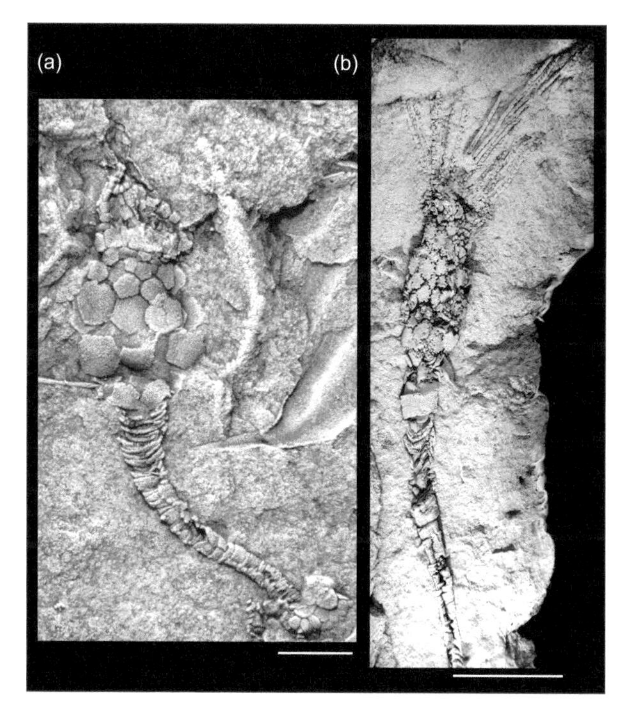

Figure 7 Gogiid eocrinoids develop epispires in later ontogenetic stages. (a) Juvenile specimen of *Akadocrinus jani* (latex cast of L42227a); like many other juvenile taxa of gogiid eocrinoids (under ~3 mm in height), this specimen has no epispires. (b) An older specimen of *Akadocrinus jani* (L42222), where epispires have developed along the plate sutures as is typical in epispire-bearing eocrinoids. Both are modified from Nohejlová and Fatka (2016). Specimens whitened with ammonium chloride sublimated. Scale bar 1 = 1 mm; 2 = 10 mm.

structures in adults (Fig. 7(b)) (McKinney and Sumrall, 2011). Epispires first developed near the top portion of the theca, toward the ambulacral area, then along the bottommost portion of the theca by the stem. Parsley (2013) found that by the time thecae reached about 8 mm in height, epispires were distributed evenly across the body. Parsley (2013) noted that the emergence of the epispires in this particular pattern suggests that they formed to support high levels of metabolic activity. Some eocrinoids have lost epispires throughout all of their ontogenetic stages; in such cases, taxa without epispires typically had much thinner plates (and often with flattened thecae, like *Haimacystis*) than those with epispires (Sumrall et al., 2001; Sheffield et al., 2022). Presumably, with thinner plates came a reduced need for pores to perform respiratory function; organisms would have been able to respire across the entire thecal surface (Sprinkle, 1975). Well-documented and complicated developmental pathways for respiratory structures in blastozoans make it difficult to ascribe plate series to perforate or imperforate extraxial skeleton using a strict diagnosis based on the presence of pores within the EAT framework. Again, it is the historical context of the plate series' origin that determines the homology of the plating, not the presence or absence of pore systems.

There are several examples of echinoderm taxa that have evolutionarily lost their respiratory pore structures or have significantly reduced them. Glyptocystitid rhombiferans are a clade that bear pectinirhombs among derived members, but plesiomorphic taxa either lack respiratory structures or bear a variety of other respiratory structures (Paul, 1968a; Sprinkle and Wahlman, 1994). Pectinirhombs are added to specific plate sutures ontogenetically and dichopores are sequentially added to existing pectinirhombs ontogenetically (Paul, 1968b; Sumrall and Sprinkle, 1999; Sumrall and Schumacher, 2002). *Amecystis* is derived within the pectinirhomb-bearing pleurocystitid clade (Paul, 1967; Parsley, 1970; Broadhead and Strimple, 1975; Sumrall and Sprinkle, 1995), but lost its respiratory structures, something that has happened more than once in the pectinirhomb-bearing rhombiferans. To respire, *Ameystis* likely utilized a mode of respiration that some modern holothuroids use, cloacal pumping via a large, flexible integument of the periproct (Broadhead and Strimple, 1975). Other Paleozoic echinoderm taxa, such as some edrioasteroids, may have also used modes of cloacal pumping to respire (see Bell, 1976b). In glyptocystitoids, the ephemeral nature of respiratory structures shows that either many taxa with perforated extraxial skeleton lack pores or many with imperforate extraxial skeleton bear pores.

In some cases, respiratory pore systems are induced in plate series, regardless of their origin, showing that functional constraint is independent of the homology of the particular elements. For example, asteroblastid diploporans have diplopores that are restricted to the interambulacral areas (Kesling, 1968).

In other cases, diplopores can be present in both axial and extraxial skeletal elements, such as in *Tristomiacystis*, a Devonian diploporan (Sumrall et al., 2009). This taxon bears diplopores on both the thecal wall plates (extraxial), as is typical for diplopore-bearing taxa, as well as diplopores piercing the floor plates (axial). Note that the floor plates follow OPR, bear the food groove and brachiole facets, structurally form the thecal wall without underlying thecal plates, and conform to all morphological and developmental expectations of abradial floor plates as described by Sumrall (2017). In essence, using a strict diagnosis, these plates can be both axial and extraxial. In addition, taxa such as *Dactylocystis*, *Revalocystis*, and *Estonocystis* have a reduced number of diplopores and constrained their placement to the axial ambulacral floor plates (Fig. 1 and Fig. 8) (Kesling, 1968; Sheffield and Sumrall, 2019). Chauvel (1941) suggested that constraining diplopores to the ambulacral floor plates could indicate that at least some respiratory structures may have had an ambulacral origin. While testing that hypothesis is not within the scope of this Element, this comment predicts that in some lineages, respiratory structures might begin in the axial system and migrate toward the extraxial plate series. However, we do not see that in blastozoan morphology. A clear example of this can be seen in glyptocystitoids, which show a disparate range of respiratory structures: those with corrugated thecal plates, those with epispires, those with pectinirhombs, and some without any respiratory structures at all (Sheffield et al., 2022). In all these examples, these disparate respiratory structures evolved independently and appear solely in the extraxial skeleton.

One aspect of many categories of respiratory pores is the fact that they do not penetrate the thecal wall. Endothecal respiratory structures such as dichopores, cryptopores, and hydrospires have incurrent and excurrent pores, but the entirety of the respiratory structures is contained within the thecal wall. Thin folds on the thecal interior facilitate gas exchange through the porous stereom but ambient seawater is fully external to the theca. Conversely, some exothecal respiratory structures such as humatirhombs and some forms of diplopores have the entire pore system contained within the thecal plate but differ in that the coelomic fluid circulates through the pore system. In none of these cases are the pores truly perforations. Epispires do perforate the thecal wall, but this is only true because the papulae that presumably pass through these structures are noncalcifying.

3.1.2 Hydropore, Gonopore, Periproct

Extraxial–axial theory describes oral plates and oral frame plates as part of the extraxial skeleton based on the hydropore, gonopore, and periproct commonly perforating these plate series. This is the derived condition for these plates, as

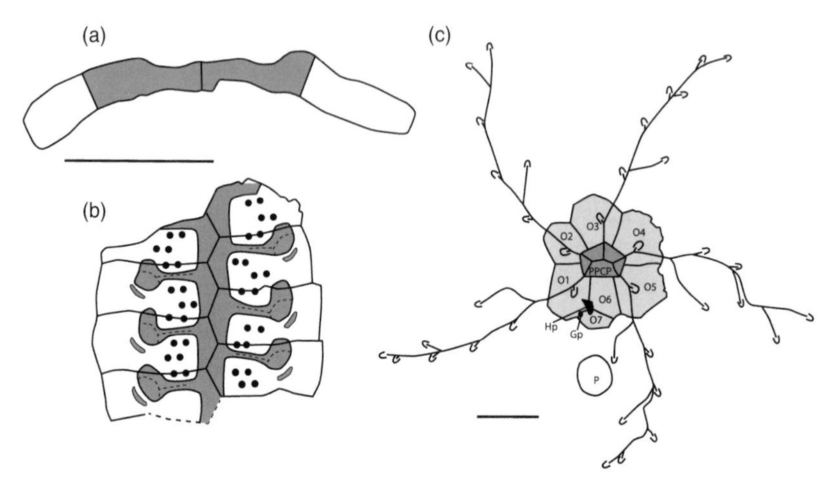

Figure 8 Aberrant morphologies within diploporans. (a) Line drawing of cross section of an ambulacrum of *Dactylocystis mickwitzi* (PIN 17186); ambulacral floor plates colored gray. (b) Line drawing of an ambulacrum of *Dactylocystis mickwitzi* (PIN 17186); brachiole facets connect to main ambulacral groove and rest upon ambulacral floor plates. Axial ambulacral floor plates are pierced by diplopores. (c) Line drawing of the oral area and ambulacra of *Glyptosphaerites leuchenbergi* (PIN 17172); the ambulacra lie directly against the theca without underlying floor plates. The ambulacra extend down the theca in the absence of axial skeleton. Scale bars = 5 mm.

can be seen by their position among early taxa such as *Lepidocystis* and *Kailidiscus* (Sprinkle, 1973; Zhao et al., 2010). As described in the following, homology is defined by development rather than diagnosed by the features of skeletal elements. From this perspective, oral plates and oral frame plates have been clearly established as part of the axial skeleton, as they are developmentally the earliest formed portions of the floor plate series and they have been shown to be morphologically contiguous with the abradial and adradial floor plate series respectively (Zhao et al., 2010).

Consequently, the placement of the hydropore, gonopore, and periproct is not restricted to the perforate extraxial skeleton. These features show heterotopic evolution, resulting in a variety of positions within the theca, including placement in both the axial and extraxial skeleton as well as the sutures between them. In several early taxa, such as imbricate and gogiid eocrinoids and *Kailidiscus*, the hydropore and gonopore are in the form of small pyramids positioned in the proximal right CD interambulacrum within the extraxial skeleton of the interambulacral plating and the periproct is positioned centrally to distally in the CD interambulacrum (Sprinkle, 1973; Zhao et al.,

2010). This is consistent with its traditional diagnosis as perforate extraxial skeleton; many of these early groups bear respiratory structures in these plates, though there are exceptions, such as *Kailidiscus* and some gogiids. In more derived taxa bearing oral plates (e.g., derived blastozoans and edrioasterids), the hydropore and gonopore are nearly universally positioned within the axial skeleton, namely, the oral plates and the oral frame plates; typically, this is seen in the posterior oral plate series shared between O1, O6, and O7 (Paul, 1968a; Sumrall and Wray, 2007; Sumrall and Waters, 2012; Kammer et al., 2013).

In isorophid edrioasteroids that bear oral frame plates, the hydropore and gonopore are incorporated into the oral frame (Kesling and Mintz, 1960; Bell, 1976b; Sumrall, 1996), and this combined hydrogonopore is covered by one or a series of hydropore orals (Kesling, 1960; Bell, 1976b; Sumrall, 1996). These plates appear to be a combination of modified ambulacral cover plates and interambulacral plates resulting in the orifice bordering the axial and extraxial skeleton externally and positioned in the axial skeleton internally. In *Euryeschatia*, the hydrogonopore is bordered between the hydropore oral, repositioned along the C ambulacrum in the CD interambulacrum and bordered internally by ambulacral floor plates (Sumrall and Zamora, 2012).

Similarly, the positioning of the periproct is not limited to the perforate extraxial skeleton. While most early echinoderms bear the periproct in the extraxial skeleton, typically in the interambulacral plating of the CD interray, this is not universally the case. Some taxa have moved the periproct to the side of the theca with the plating of the thecal wall into what is typically inferred to be perforate skeleton, which is seen in derived glyptocystitoid rhombiferans and paracrinoids (Parsley and Mintz, 1975; Sumrall and Waters, 2012; Zamora et al., 2017). Several cases in later taxa also exist where the periproct is positioned more proximally and can border the oral plate series, such as the stem rhombiferan, *Ridersia*, blastoids, and the diploporan, *Tristomiacystis* (Sumrall et al., 2009; Zamora et al., 2017). Still others, including some eublastoids, have the periproct completely bordered by oral plates. The position of these orifices, which are functionally quite different, is, again, not an indication of plate series homology, but a reflection of heterotopy in the evolutionary history of these clades.

Furthermore, the developmental origin of the hydropore, which has been classified as part of the perforate extraxial skeleton, varies across extant taxa. During the development of the echinoid *Paracentrotus lividus*, the calcified madreporite found in the adult body plan forms around the larval hydropore as a result of further biomineral deposition in continuity with the larval skeleton (Gosselin and Jangoux, 1998). In this echinoid, the adult hydropore is thus the same structure as the larval hydropore. In the asteroid *Asterias rubens*, however, the hydropore of the larvae is closed following metamorphosis, and the madreporitic pore arises as a distinct canal connected to the coelomic cavity. Thus, the larval hydropore and adult madreporite are distinct structures (Gondolf, 2002). This suggests that the developmental origin of adult structures, such as the madreporitic pore from the larval hydropore, is not consistent across different echinoderm groups.

The plates that bear or support the gonopore have also been classified as perforate extraxial skeleton (Mooi et al., 1994). The developmental origin of the gonopores, however, is vastly different from that of the hydropore. In contrast to the hydropore, the gonopores do not have an analogous structure in the larva. In echinoids, the gonopores do not open until sexual maturity (Spirlet et al., 1994). Prior to this, the gonopore-bearing plates in most echinoids, the genital plates, lack any perforation.

Furthermore, the perforations associated with the gonopores are not only limited to the genital plates of crown group echinoids. In some derived holasteroid irregular echinoids, as well as the tithoniids (stem group atelostomates), some of the gonopore openings are found on ocular plates instead of genital plates (Saucède et al., 2001; Smith, 2004; Gaillard et al., 2011). Additionally, in some clypeasteroid echinoids, as well as the Cretaceous atelostomate *Absurdaster*, the gonopore opens within the interambulacral plating (Kier, 1968; Kroh et al., 2014), interpreted by (Mooi et al., 1994) to be axial skeleton. The disparate developmental origins found among perforations in different plate types, and the migration of perforations such as the gonopore across different, nonhomologous plate types suggest that the perforation of skeletal elements may be unrelated to homology, and may instead be associated with signaling from the gonoduct (which connects the gonad to the genital pore) to the skeleton, which subsequently results in the local resorption of the skeleton and a resulting perforation of the skeleton.

Okada (1979) surgically removed portions of the gonoduct from juvenile and young postjuvenile echinoids and found that removal of the gonoduct midway through perforation of the genital plate resulted in the further cessation of genital pore formation. Furthermore, Okada found that when the gonad was removed, occasionally an additional gonoduct would regenerate from it and pierce an interambulacral plate, resulting in a new gonopore forming in the interambulacra. Likewise, when the gonoduct was removed and regeneration resulted in the development of multiple gonoducts, a gonopore formed where each contacted the skeleton. All of this suggests that the perforate nature of the genital plate has little to do with the innate ability of this skeletal element to be perforate, but with a resorption-related signal sent from the gonoduct.

3.2 Feeding Structures

The feeding structures of echinoderms, ambulacra, are made up of three plate series with, in some cases, accessory appendages (Sumrall, 2017). The food groove is floored by a series of plates called floor plates that form two series distributed among taxa. The food groove is covered by a series of ambulacral cover plates. Distinguishing the floor plate series is complex but can be framed in the form of a testable hypothesis. If both plate series are present as in pyrgocystids and *Kailidiscus* (Zhao et al., 2010; Sumrall and Zamora, 2011, 2018), position can be used with the adradial set along the ambulacral midline and the abradial set offset from the midline. Where known, the adradial set is wholly internal and only seen in interior views of the theca and cross sections. The abradial set is typically expressed with a broad laterally exposed shelf abradial from the cover plate series (Sumrall, 2017).

Hypotheses of floor plate homology can also be tested because the floor plate series and the peristomial border series are developmentally linked and show a consistent distribution among taxa. Taxa bearing oral frame plates possess adradial floor plates, and this floor plate set forms a series extending from and morphologically contiguous with the oral frame plates proximally. Taxa bearing oral plates bear abradial floor plates that begin with the oral plates proximally (Sumrall, 2017; Zamora and Sumrall, in press). There are, however, some cases where floor plate homologies are ambiguous or absent. It has been convincingly shown that many early crinoids bear floor plates within the structure of the erect ambulacral system (Guensburg et al., 2010). But whether these taxa bear adradial or abradial floor plates remains unknown. The presence of an oral plate series in some early crinoids suggests that they would correlate to abradial floor plates, but other data are lacking (Sumrall, 2017).

Interestingly, while early crinoids almost universally have uniserial brachial elements, many derived crinoids have biserial brachial growth that identically patterns plate series following the OPR. In these cases, the brachial elements are interpreted as extraxial simply based on phylogenetic arguments despite the fact that the OPR would diagnose them as axial. A further complication is that podial pores are present in many early taxa and eleutherozoans. These pores are intimately associated with the water vascular system and presumably follow the trace of a radial canal (Guensburg et al., 2020). Pores can be present in either or both adradial and abradial floor plates and in some cases along sutures between them as in *Kailidiscus*. However, the most prominent pattern is the reduction of pores to podial basins or more typically a complete loss of these pore systems (Guensburg et al., 2020). This is not to say that the water vascular system and podia are missing entirely, *sensu* Sprinkle (1973), only that the skeletal evidence for podia is lacking. In any case, floor plate systems can be convincingly shown to be homologous regardless of the presence or absence of podial pore systems.

In some cases, as in a number of sphaeronitid diploporans (Sheffield and Sumrall, 2019), the food grooves do not lie upon the floor plates, but rather directly upon the plates of the body wall. In these cases, the food groove does not directly follow plate sutures, nor plate positions but rather extends down the theca in the absence of axial skeleton distal to the peristomial border, except for cover plates. In addition, small side food grooves lead to elevated brachiole facets that are induced as out swellings of the thecal plates themselves, rather than being borne on floor plates. The lack of any patterning to the plates bearing the food groove such as terminal growth and the inconsistency in plating ray to ray and specimen to specimen argue against the presence of floor plates. Many of these examples have a 36° rotation of the ambulacral system with respect to the underlying oral plate configuration. Presumably, cover plates (axial skeleton) would have articulated directly with these thecal plates, though no direct evidence has been observed, as these plates are taphonomically the least likely to be preserved (Brett et al., 1997). Because the soft tissue component of ambulacra extends along these structures, it is likely that the floor plate component of the axial skeleton is simply not calcified. Although brachioles are not preserved in these taxa, we can assume that if they are calcified, they are also part of the axial skeleton. This is seen in several diploporan taxa such as *Glyptosphaerites* and most sphaeronitids (Paul (1984) and Paul and Toom (2021) suggest that *Glyptosphaerites* is a sphaeronitid; however, that relationship was not uncovered during phylogenetic analyses by Sheffield and Sumrall (2019)).

In Eublastoidea, the location of the food groove is variable across taxa. In some taxa such as *Cryptoschisma*, *Pentremites*, and *Deltoblastus*, once the primary leaves the oral plate suture, the food groove is borne axially along the lancet plate, which is of thecal origin (i.e., extraxial skeleton; Fig. 9(a)) (Sumrall, 2017). From this main food groove, side food grooves arise, which lead to pairs of floor plates (often referred to as side plates), which bear brachiole facets. This is a derived condition, because more plesiomorphic taxa, such as *Troosticrinus* and *Hyperoblastus*, show the primary food groove lying upon the periradial suture of the biserial floor plates (Fig. 9(b)). In essence, the lancet evolutionarily erupted through the periradial suture and took on the role of bearing the primary food groove in the more derived taxa. In some taxa, the lancet plate is not fully exposed and the primary food grove rests on both the lancet and floor plate series (Sumrall and Waters, 2012).

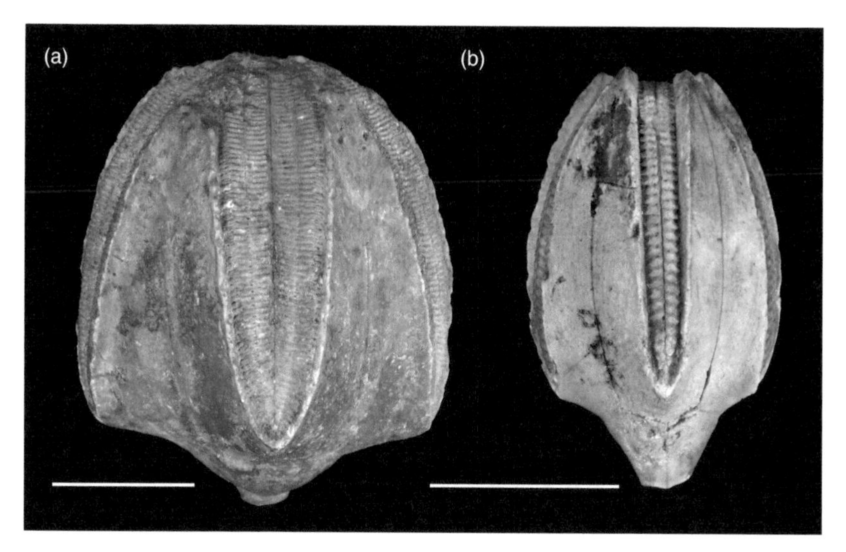

Figure 9 Examples of eublastoid specimens with variable food groove placement due to exposure or lack thereof of the lancet plate. (a) *Pentremites cervinus* (UMMP 1418) specimen with the food groove situated on the lancet plate. (b) *Hyperoblastus alveata* (UMMP 37809) specimen with the food groove placed on the floor plates. Scale bar = 10 mm. Image provided by the University of Michigan Museum of Paleontology under a CC-BY-NC 4.0 and published with permission here.

The preceding examples show that skeletal regions within blastozoans do not simply fit general diagnoses for plate types. Great care must be taken when assigning skeletal homology based on ephemeral features, and in no case is the distinction more dubious than differentiating perforate and imperforate skeleton. Perforate skeleton can be imperforate, ambulacral floor plates can be perforated through respiratory structures, and pores are associated with podia, both of which are absent in the majority of taxa.

3.3 Oral Surface Plating

The identification of two series of peristomial bordering plates as distinct plate series and part of the axial skeleton is well founded (Bell, 1976a; Sumrall and Waters, 2012). Typically radially positioned oral frame plates and interradially positioned oral plates are present in edrioasterid edrioasteroids, and their precursor plate series are both present in *Kailidiscus* (Fig. 3) (Bell, 1976a; Zhao et al., 2010). In *Kailidiscus*, the oral frame plates are represented by a series of pore bearing elements that form the immediate peristomial border and are continuous with (though highly modified from) the adradial floor plate series (Fig. 3). In later edrioasteroids, this distinction remains clear with *Edriophus* having oral frame plates that taper disto-radially and do not form a series with the abradial floor plate series that in turn forms the flood groove (Bell, 1976a; Sumrall and Waters, 2012). In isorophids, the oral frame plates form a continuous series with the adradial floor plates. The podial pore-bearing oral plates in *Kailidiscus* similarly form an unbroken plate series with the abradial floor plate series, which in most respects is identical to the plating arrangement in *Edriophus* where these plates form an unbroken series (Bell, 1976a; Sumrall and Waters, 2012).

While some authors do not make this distinction, either counting the number of plates around the peristome or using nonhomologous naming schemes (which can impact downstream analyses; for reference, see Bauer et al., 2022), such activities confuse the literature and mislead phylogenetic analyses. In reality, the distinction between oral plate and oral frame plate is clear and unambiguous in nearly every case. There are a few examples that are more complex and require more detailed analysis, such as the case of the diploporans with the oral frame plate shift discussed in Section 3.2. While the radial position of these elements is consistent with oral frame plates, other features unambiguously show them to be oral plates. First, they show the typical broadly exposed external shelf beyond the cover plates, whereas oral frame plates are strictly internal. Second, the presence of seven plates with the correct distribution among the ambulacra and the positioning of the hydropore and gonopore shared among the O1, O6, O7 complex is consistent with other taxa bearing oral plates. This same situation is present in holocystitid diploporans (Fig. 10), where a series of differentiated proximal thecal plates take on ambulacral function (periorals) but are not part of the homologous oral frame series (Sheffield and Sumrall, 2017, 2019). Indeed, these plates are distal to the oral plate series that have unambiguous plate homologies.

3.4 Water Vascular System

The first components of the water vascular system form in the rudiment during the earliest developmental phases of the adult echinoderm body plan. The rudiment is derived from the left coelomic pouch in indirect developing echinoderms and is the first structure to show the pentaradial symmetry present among members of the crown group. As the rudiment develops, the left hydrocoel forms in the characteristic pentaradial shape (Peterson et al., 2000; Mooi and David, 2008; Morris, 2012), which develops into the primary podia seen in juvenile echinoids and at the distal end of the arm in ophiuroids and asteroids. These primary podia form prior to metamorphosis and are lost or resorbed in some echinoids (Thompson et al., 2021).

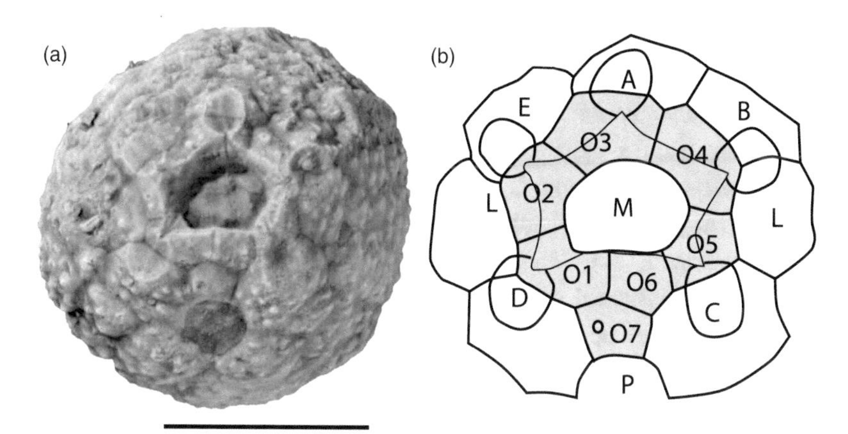

Figure 10 Rotation of ambulacra upon the oral surface. (a) Oral view of *Holocystites scutellus* (SUI 48183). (b) Line drawing of the same specimen, bearing a 36° rotation of the ambulacra upon the oral surface. The five ambulacral grooves lie against the middle of the oral plate series, as opposed to lying on the sutures between the oral plates; the latter is the more commonly seen condition in blastozoans. O = oral plates; L = lateral nonfacetal bearing plates; M = mouth; P = periproct; A–E = ambulacra. Both are modified from Sheffield and Sumrall (2017). Specimen whitened with ammonium chloride sublimated. Scale bar for 1 = 10 mm.

After the initial development of the water vascular system, new podia are added to either side of the radial water vessel in an alternating metameric manner near the aboral end of each ray (Morris, 2007; Formery et al., 2021). The development of new podia is underlain by proliferation of mesodermal cells (Thompson et al., 2021), and in echinoids, the formation of these new podia corresponds with the addition of new overlying ambulacral plates (Gao et al., 2015; Thompson et al., 2021). In juvenile echinoids, each podium protrudes through a single ambulacral pore in each plate, which, at early postmetamorphic stages of growth, lacks the interporal partition that characterizes echinoid pore pairs. In at least some taxa, the pore through which the podia protrude is at the border between two sequentially added plates (Gosselin and Jangoux, 1998; Gao et al., 2015). This is also similar to the podia that span across multiple ambulacral plates seen in some bothriocidaroid echinozoans (Thompson et al., 2022), and is also reminiscent of the pores in *Kailidiscus* and the shared podial basins of asterozoans (Zhao et al., 2010).

In echinoids, the pores through which the podia protrude form coincident with development of the podia, as the new plate is added marginal to the apical disk (Gao et al., 2015). The formation of pores through which the podia protrude is suggestive of a signaling mechanism from the podia to the skeleton of the ambulacral plating, which may induce the opening of the pore. Though it is only speculation, similar signaling mechanisms may underlie the existence of multiple disparate and nonhomologous perforations across the plating of different echinoderm groups, particularly those perforations which are housed in the perforate extraxial skeleton.

Another unanswered question concerns the nature of skeletal types when plate series are decalcified (Zamora et al., 2022). From one perspective, if the skeleton (e.g., the floor plates in *Glyptosphaerites*) is not present, then there is no axial skeleton in the floor plate series. However, it may also be the case that the soft mesodermal tissue that is responsible for biomineral deposition is still present in the organism, though taphonomically not preserved. Both of these cases would have an identical expression in the fossil record but would result from fundamentally different developmental origins.

4 Reconciling EAT and UEH

4.1 Blastoids and Hemicosmitoids as Examples

To show how these two homology schemes can be used in concert, a comparison is made between the blastoid *Pentremites* and the hemicosmitoid *Hemicosmites*. Both taxa have axial skeletal elements associated with the ambulacral system consisting of oral plates forming the peristomial border (orals in hemicosmitids, deltoids in *Pentremites*). From these, double biserial floor plates extend along the radii. In blastoids these are the side plates running along the edge of the ambulacra, and in *Hemicosmites* these are erect ambulacra that mount onto paired plates incorporated into the oral area inferred to be the first pair of flooring plates (Sumrall, 2010; see Paul, 2021 for another interpretation). Biserial brachioles articulate to facets born on the sutures between primary and secondary floor plates. *Pentremites* has an unusual situation where a radially positioned plate, the lancet (see what follows), erupts developmentally through the perradial suture and bears the main food groove and medial portions of the side ambulacra. Sutures between paired primary and secondary floor plates form facets from which biserial brachioles arise.

In both taxa, primary peristomal cover plates cover the peristome and grade into the proximal portions of the ambulacral cover plates (Sumrall and Waters, 2012). These are only distinguishable in the earliest ontogenetic stages in *Pentremites* (passalocrinid stage) and are poorly documented in *Hemicosmites* (Bockelie, 1979). Interestingly, in *Pentremites*, the ambulacral cover plates along the main food groove transition from the axial skeleton along the oral plate sutures, transition onto the extraxial skeleton (lancet plate), and then transition back onto the axial skeleton (side plates). These cover plates continue onto the brachioles covering up the distal-most portions of the food groove (Sumrall and Waters, 2012).

In *Hemicosmites*, cover plates extend along the erect ambulacra, where in some taxa they are greatly enlarged (Sprinkle, 1975; Sumrall et al., 2015) but are a simple biseries in other taxa. Cover plates then extend up the brachioles covering the distal-most food grooves.

The extraxial skeleton is divided into two regions: the theca, which is generally interpreted to be perforate extraxial, and the stem, which is generally interpreted as imperforate extraxial. In *Hemicosmites*, the thecal surface is covered by endothecal respiratory structures in the form of cryptorhombs. Furthermore, the periproct perforates this skeleton, suggesting that the thecal plating is perforate extraxial skeleton. In *Pentremites*, the situation is more complex. Incurrent pores are positioned between the floor plates and the thecal wall. In the deltoid region, this is axial skeleton of the oral plate and in the radial region it is extraxial (presumably perforate). The excurrent pores exit through gaps between a combination of the oral plates (axial), floor plates (axial), and lancet plates (extraxial). The periproct is positioned similarly, as it is a combination anal opening gonopore and respiratory structure. In both taxa, the stem is inferred to be imperforate extraxial skeleton, as it is in other blastozoan taxa

4.2 Taxa That Are Difficult to Reconcile

As new fossil discoveries are made, we have to continuously reevaluate our working hypotheses by incorporating new evidence. Echinoderms are extremely disparate and host a vast array of skeletal morphologies and bauplans (Deline et al., 2020). It is not surprising that new fossil finds continue to challenge the existing hypotheses we construct to understand them more completely. For example, the discovery of a *Tholocystis* specimen from Katian-age deposits of Sardinia (Sumrall et al., 2015) challenges reconciliation with the UEH hypothesis. The specimen has clear diplopores piercing the thecal plates, placing it within the broader diploporan group, but further reconciliation within the UEH framework is difficult, as complete specimens preserving the full suite of axial skeletal elements in the oral area have yet to be found. The Sardinian *Tholocystis* specimen (Fig. 11(a)) has unusual ambulacra that are wide and recumbent against the theca, but details concerning the peristomial border, number of oral plates, position of the hydropore gonopore and periproct, and the nature of the ambulacral floor plates (or potential lack thereof) make it difficult to interpret at this time (Sheffield and Sumrall, 2019).

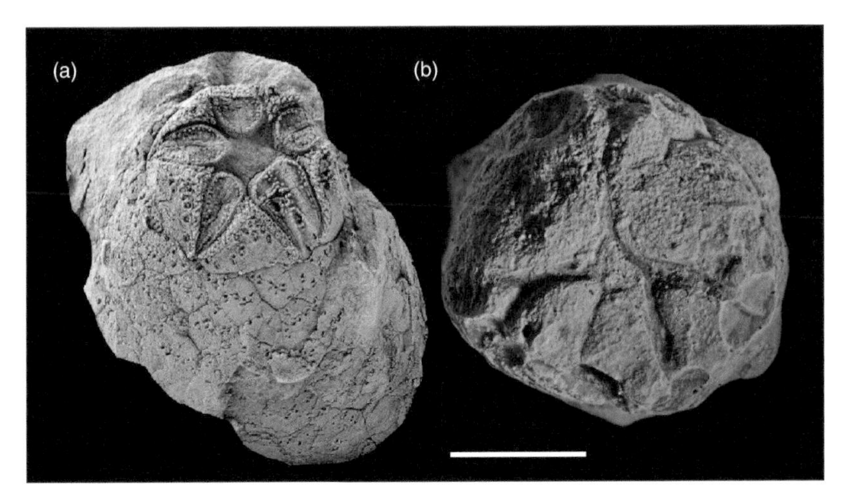

Figure 11 Taxa that have posed challenges to echinoderm homology schemes. (a) Oral view of latex cast of *Tholocystis* sp. (MGM-7192-X; modified from Sheffield and Sumrall [2019]). *Tholocystis* is known from incomplete oral areas, and in this specimen, the missing oral plates, ambulacral floor plates, hydropore, and gonopore make it difficult to interpret within a UEH framework. (b) Oral view of *Thomacystis tuberculata* (BMNH E16300). Previous morphological interpretations have been published that state it bears four ambulacra; however, this specimen bears five ambulacra in the plesiomorphic 2–1–2 condition and a normal arrangement of oral plates. Specimens whitened with ammonium chloride sublimated. Scale bar = 1 cm.

Another taxon whose morphology is not immediately understandable through the UEH framework is the hemicosmitoid *Thomacystis* (Fig. 11(b)). *Thomacystis* appears to bear four ambulacra, presumably with the A ambulacrum reduced and B–E being present, and one or two erect ambulacra (maybe four), arising from plates bordering the mouth (Paul, 1984; Sumrall and Waters, 2012; Paul, 2021). The confusion arises because taxa that bear this ambulacral arrangement have a non–ambulacrum-bearing suture between O3 and O4, *Thomacystis* has a single plate in this position. This condition would be highly derived from other hemicosmitoid rhombiferans and likely to represent apomorphic features. Paul (2021) also noted that there are questions when attempting to reconcile the morphology of *Thomacystis* within an EAT framework.

However, another interpretation shows it to have five ambulacra in the normal 2–1–2 configuration with extremely long shared ambulacra and short B–E ambulacra (Fig. 11(b)). The unusual morphology is not in the loss of the A ambulacrum, but in the relatively long shared ambulacra, which is typical for hemicosmitoids. While *Thomacystis* needs further study to clarify these issues, it highlights one of the greatest challenges in echinoderm phylogenetics: recognizing the difference between variations on a theme and truly novel morphologies.

Although tegmens of monobathrid crinoids are easily reconcilable within the UEH framework, showing clear vestiges of the PPCPs and ambulacral cover plates, diplobathrid crinoids are not, as they lack an easily recognizable organization (Kammer and Ausich, 2007; Kammer et al., 2013). There are two possibilities to explain this. First, it is possible that the tegmens of diplobathrids bear the cover plate elements as suggested by the UEH hypothesis, but they lose distinctiveness later in ontogeny as new plates are added with increased maturity and sutural relationships are modified to accommodate these additions. Indeed, loss of PPCP differentiation by size ontogenetically is common in many edrioasteroid taxa but they can be identified by their constrained position. Ultimately the condition in diplobathrids is testable if earliest ontogenetic stages can be seen in which the PPCPs and ambulacral cover plates retain the plesiomorphic morphology. The second possibility is that the diplobathrid tegmen is a novel feature and unrelated to the plesiomorphic ambulacral system.

5 Future Areas of Study

Extraxial–axial theory and UEH as homology hypotheses do not exist in opposition to one another. They address different aspects of skeletal homology among echinoderms. However, it must be remembered that these are homology hypotheses and as such must be continuously tested and refined as part of the scientific process. This means that these hypotheses of homology must be continually evaluated via discovery of new taxa, reinterpretation of known taxa as more complete material comes to light, and through the utilization of rigorous phylogenetic methods, as opposed to relying solely on expert opinions of which features are group defining (Wright, 2015). Ultimately, the understanding of phylogenetic relationships rests upon our understanding of homology and how it is transformed through the evolutionary process. It is therefore imperative to continue moving forward along these lines of research so that a more complete picture of these animals can be achieved.

There must also be recognition that the current typological definitions of axial and extraxial are too strict, as they diagnose skeletal type based on features rather than define them based on evolutionary history. Examples discussed in this Element document that Paleozoic echinoderms show broader variation than what is captured in the diagnoses centered around OPR = “axial” and structures piercing the thecal wall = “perforate extraxial.” This does not mean that these categories are not helpful in understanding the evolution and development of echinoderms; rather, we must continue to study the variation and development of the axial and extraxial skeletons outside of Eleutherozoa to ensure that they can be applied with high fidelity across noneleutherozoan groups. Certainly, new echinoderm fossils that challenge our standing hypotheses of homology will be found, such as the examples of *Thomacystis* and *Tholocystis* discussed earlier.

We must also be cognizant that exceptions to rules do not mean that the ideas behind them are fallacious. Simply stating that tetrapods have four legs would deny group membership of the numerous clades of tetrapods that lack limbs. Exceptions simply highlight areas that need further study. New data challenging our present knowledge is the very nature of science; we must continue to be willing to question what we consider to be true and our assumptions to move toward a more accurate understanding of the evolutionary history of pan-Echinodermata.

However, the preceding examples do not indicate that EAT and UEH are poor tools to understand homology. All models are wrong, but some are useful (Box, 1976). We would contend that EAT and UEH are both wrong to some degree, but both are useful. Extraxial–axial theory is a powerful tool for defining characters of the gross morphology of the theca dividing it along developmental and therefore homologous lines. Universal elemental homology is a powerful tool for understanding the details of the axial skeleton, similarly, defined along developmental lines to recognize homology. These tools, when applied with care and evidence, place character construction into the framework of testable hypotheses and generate interpretations that are internally consistent across a dataset such that character data can be coded, analyzed, and refined. The fossil record will continue to provide new challenges and we must continuously test and refine our hypotheses, and these tools will aid us in this endeavor. As we continue to learn more about the breadth of diversity in the fossil record and combine it with new advances in understanding the development of the echinoderm system, we can begin to build toward a grand, unified hypothesis of echinoderm homology.

The wide applicability of using growth zones as a means of establishing homology of divergent features across various echinoderm groups merits further investigation. An exciting avenue of new research within the EAT framework would be to build on previous work using growth zones to establish homology across divergent features. This could be particularly useful in attempting to establish grounds for potential homology of features across divergent echinoderm groups, such as both pentaradial and asymmetric forms. There currently is no consensus regarding the homology of the ambulacra in radiate forms to the skeletal plates of bilaterally symmetrical or asymmetric fossil echinoderms. Precise and detailed analyses of ontogeny and mode of plate addition, either via a distinct growth zone or via intercalation, may help to clarify some of these issues.

Another avenue of fruitful research would be to investigate the genetic underpinnings of development in extinct clades. In many animal groups, expression patterns of different genes and components of genetic regulatory networks are used as a basis for establishing homology of morphological characters across wide phylogenetic distances (e.g., Tweedt, 2017). This work has been particularly well developed in studies of arthropods and other ecdysozoan phyla (e.g., Ortega-Hernández et al., 2017; Janssen and Budd, 2020; Lev et al., 2022). Despite the long history of work attempting to understand gene expression during development in echinoderms, studies within the phylum have lagged behind those in other animal groups with regard to the use of molecular tools to establish homology. Much effort has been invested in the last 20 years attempting to understand the expression patterns of *Hox* genes and other homeodomain-bearing transcription factors during development of the echinoderm adult body plan (e.g., Arenas-Mena et al., 2000; Morris and Byrne, 2005; Hara et al., 2006; Cisternas and Byrne, 2009; Morris and Byrne, 2014; Tsuchimoto and Yamaguchi, 2014; Kikuchi et al., 2015; Byrne et al., 2016). These *Hox* genes are often expressed sequentially along the antero-posterior or oral–aboral axis of divergent echinoderm classes and/or in the coelomic cavities. Despite this excellent work, there is little evidence that any of these genes are involved in development or patterning of the skeleton and are often expressed in cells distinct from those expressing skeletogenic markers (Tsuchimoto and Yamaguchi, 2014).

A future fruitful avenue for research will be to identify differential gene expression in the development of different features of the adult echinoderm skeleton. Initial work has been done surveying several transcription factors and differentiation genes that are expressed during growth and regeneration of different skeletal structures in the arm of the brittle star *Amphiura filiformis* (Piovani et al., 2021). This work suggests that different combinations of skeletal genes are responsible for the development of different skeletal structures. With further data from different echinoderm classes, comparative analyses (e.g., (Erkenbrack and Thompson, 2019)) will provide a framework for understanding potential homology of the morphologically diverse skeletal structures seen in different echinoderm groups.

References

Arenas-Mena, C., Cameron, A. R., and Davidson, E. H. (2000). Spatial expression of Hox cluster genes in the ontogeny of a sea urchin. *Development*, **127**, 4631–4643.

Ausich, W. I., and Kammer, T. (2013). Mississippian crinoid biodiversity, biogeography and macroevolution. *Paleontology*, **56**, 727–740.

Ausich, W. I., Wright, D. F., Cole, S. R., and Sevastopulo, G. D. (2020). Homology of posterior interray plates in crinoids: A review and new perspectives from phylogenetics, the fossil record and development. *Palaeontology*, **63**, 525–545.

Bauer, J. E. (2020). Paleobiogeography, paleoecology, diversity, and speciation patterns in the Eublastoidea (Blastozoa: Echinodermata). *Paleobiology*, **47**, 221–235.

Bauer, J. E., Sheffield, S. L., Sumrall, C. D., and Waters, J. A. (2022). Echinoderm model systems, homology, and phylogenetic inference: Comment and reply to Paul (2021). *Acta Palaeontologica Polonica*, **67**, 455–468. DOI: https://doi.org/10.4202/app.00956.2021.

Bell, B. M. (1976a). A study of North American Edrioasteroidea. *New York State Museum and Science Survey*, **21**, 1–447.

Bell, B. M. (1976b). Phylogenetic implications of ontogenetic development in the class Edrioasteroidea (Echinodermata). *Journal of Paleontology*, **50**, 1001–1019.

Bell, B. M., and Petersen, M. S. (1976). An edrioasteroid from the Guilmette Formation at Wendover, Utah-Nevada. *Journal of Paleontology*, **50**, 577–589.

Bockelie, J.F. (1979). Taxonomy, functional morphology and palaeoecology of the Ordovician cystoid family Hemicosmitidae. *Palaeontology* **22**, 363–406.

Box, G. E. P. (1976). Science and statistics. *Journal of the American Statistical Association*, **71**, 791–799. DOI: https://doi.org/10.1080/01621459.1976.10480949.

Brett, C. E., Moffat, H. A., and Taylor, W. L. (1997). Echinoderm taphonomy, taphofacies, and Lagerstätten. *Paleontological Society Papers*, **3**, 147–190.

Broadhead, T. W., and Strimple, H. L. (1975). Respiration in a vagrant Ordovician cystoid, *Amecystis*. *Paleobiology*, **1**, 312–319.

Brochu, C. A., and Sumrall, C. D. (2001). Phylogenetic nomenclature and paleontology. *Journal of Paleontology*, **75**, 754–757.

Brower, J. C. (1999). A new pleurocystitid rhombiferan echinoderm from the Middle Ordovician Galena Group of northern Iowa and southern Minnesota. *Journal of Paleontology*, **73**, 129–153.

Byrne, M., Martinez, P., and Morris, V. (2016). Evolution of a pentameral body plan was not linked to translocation of anterior Hox genes: The echinoderm HOX cluster revisited. *Evolution & Development*, **18**, 137–143.

Chauvel, J. (1941). Recherches sur les Cystoïdes et les Carpoïdes armoricaines. *Mémoires de la Société Géologique et Minéralogique de Bretagne*, **5**, 1–286.

Cisternas, P., and Byrne, M. (2009). Expression of Hox4 during development of the pentamerous juvenile sea star, *Parvulastra exigua*. *Development Genes and Evolution*, **219**, 613–618.

Czarkwiani, A., Dylus, D. V., Carballo, L., and Oliveri, P. (2021). FGF signaling plays similar roles in development and regeneration of the skeleton in the brittle star *Amphiura filiformis*. *Development*, **148**, dev180760.

Czarkwiani, A., Ferrario, C., Dylus, D. V., Sugni, M., and Oliveri, P. (2016). Skeletal regeneration in the brittle star *Amphiura filiformis*. *Frontiers in Zoology*, **13**, 1–17.

David, B., Lefebvre, B., Mooi, R., and Parsley, R. (2000). Are homalozoans echinoderms? An answer from the extraxial-axial theory. *Paleobiology*, **26**, 529–555.

David, B., and Mooi, R. (1998). Major events in the evolution of echinoderms viewed by the light of embryology. In R. Mooi and M. Telford, eds., *Echinoderms: San Francisco*. Rotterdam: Balkema, pp. 21–28.

Deline, B. (2021). *Echinoderm Morphological Disparity: Methods, Patterns, and Possibilities*. Elements of Paleontology. Cambridge, UK: Cambridge University Press.

Deline, B., Thompson, J. R., Smith, N. S. et al. (2020). Evolution and development at the origin of a phylum. *Current Biology*, **30**, 1–8. DOI: https://doi.org/10.1016/j.cub.2020.02.054.

Dexter, T. A., Sumrall, C. D., and McKinney, M. L. (2009). Allometric strategies for increasing respiratory surface area in the Mississippian blastoid *Pentremites*. *Lethaia*, **42**, 127–137.

Duloquin, L., Lhomond, G., and Gache, C. (2007). Localized VEGF signaling from ectoderm to mesenchyme cells controls morphogenesis of the sea urchin embryo skeleton. *Development*, **134**, 2293–2302. DOI: https://doi.org/10.1242/dev.005108.

Erkenbrack, E. M., and Thompson, J. R. (2019). Cell type phylogenetics informs the evolutionary origin of echinoderm larval skeletogenic cell identity. *Communications Biology*, **2**, 160.

Frest, T. J., Strimple, H. L., and Paul, C. R. C. (2011). The North American *Holocystites* fauna (Echinodermata: Blastozoa: Diploporita): Paleobiology and systematics. *Bulletins of American Paleontology*, **380**, 1–141.

Formery, L., Orange, F., Formery, A. et al. (2021). Neural anatomy of echinoid early juveniles and comparison of nervous system organization in echinoderms. *Journal of Comparative Neurology*, **529**, 1135–1156.

Gaillard, C., Neraudeau, D., and Thierry, J. (2011). *Tithonia oxfordiana*, a new irregular echinoid associated with Jurassic seep deposits in south-east France. *Palaeontology*, **54**, 735–752.

Gao, F., Thompson, J. R., Petsios, E. et al. (2015). Juvenile skeletogenesis in anciently diverged sea urchin clades. *Developmental Biology*, **400**, 148–158.

Gondolf, A. L. (2002). The aboral surface of *Asterias rubens* L.(Echinodermata: Asteroidea) during metamorphosis with particular reference to the hydropore and madreporite. *Invertebrate reproduction & development*, **42**, 51–59.

Gorzelak, P. (2021). *Functional Micromorphology of the Echinoderm Skeleton*. Elements of Paleontology. Cambridge, UK: Cambridge University Press.

Gosselin, P., and Jangoux, M. (1998). From competent larva to exotrophic juvenile: A morphofunctional study of the perimetamorphic period of *Paracentrotus lividus* (Echinodermata, Echinoida). *Zoomorphology*, **118**, 31–43.

Guensburg, T. E., Blake, D. B., Sprinkle, J., and Mooi, R. (2015). Crinoid ancestry without blastozoans. *Acta Palaeontologica Polonica*, **61**, 253–266.

Guensburg, T. E., Mooi, R., Sprinkle, J., David, B., and Lefebvre, B. (2010). Pelmatozoan arms from the mid-Cambrian of Australia: Bridging the gap between brachioles and brachials? Comment: There is no bridge. *Lethaia*, **43**, 432–440.

Guensburg, T. E., and Sprinkle, J. (2001). Earliest crinoids: New evidence for the origin of the dominant Paleozoic echinoderms. *Geology*, **29**, 131–134.

Guensburg, T. E., and Sprinkle, J. (2007). Phylogenetic implications of the Protocrinoida: Blastozoans are not ancestral to crinoids. *Annales de Paléontologie*, **93**, 277–290.

Guensburg, T. E., and Sprinkle, J. (2010). Emended restoration of *Titanocrinus sumralli* Guensburg and Sprinkle, 2003 (Echinodermata, Crinoidea). *Journal of Paleontology*, **84**, 566–568.

Guensburg, T. E., Sprinkle, J., Mooi, R. et al. (2020). *Athenacrinus* n. gen. and other early echinoderm taxa inform crinoid origin and arm evolution. *Journal of Paleontology*, **94**, 311–333.

Hara, Y., Yamaguchi, M., Akasaka, K. et al. (2006). Expression patterns of Hox genes in larvae of the sea lily *Metacrinus rotundus*. *Development Genes and Evolution*, **216**, 797–809.

Hotchkiss, F. H. (2012). Growth zones and extraxial-axial skeletal homologies in Asteroidea (Echinodermata). *Proceedings of the Biological Society of Washington*, **125**, 106–121.

Janssen, R., and Budd, G. E. (2020). Expression of the zinc finger transcription factor Sp6–9 in the velvet worm *Euperipatoides kanangrensis* suggests a conserved role in appendage development in Panarthropoda. *Development Genes and Evolution*, **230**, 239–245.

Kammer, T. W., and Ausich, W. I. (2007). Soft-tissue preservation of the hind gut in a new genus of cladid crinoid from the Mississippian (Visean, Asbian) at St. Andrews, Scotland. *Palaeontology*, **50**, 951–959.

Kammer, T. W., Sumrall, C. D., Zamora, S., Ausich, W. I., and Deline, B. (2013). Oral region homologies in Paleozoic crinoids and other plesiomorphic pentaradial echinoderms. *PLoS One*, **8**, e77989. DOI: https://doi.org/10.1371/journal.pone.0077989.

Kerr, A. M., and Kim, J. (2001). Phylogeny of Holothuroidea (Echinodermata) inferred from morphology. *Zoological Journal of the Linnean Society*, **133**, 63–81.

Kesling, R. V. (1960). Hydropores in edrioasteroids. *Contributions from the Museum of Paleontology, University of Michigan*, **15**, 139–192.

Kesling, R. V. (1968). Cystoids. In Moore, R. C. (ed.), *Treatise on Invertebrate Paleontology*, part S Echinodermata 1 (1): Lawrence, Kansas, and Boulder, Colorado, University of Kansas Press and Geological Society of America, pp. S85–S267.

Kesling, R. V., and Mintz, L. W. (1960). Internal structures in two edrioasteroid species, *Isorophus cincinnatiensis* (Roemer) and *Carneyella pilea* (Hall). *Contributions from the Museum of Paleontology, University of Michigan*, **15**, 315–348.

Kier, P. M. (1968). Echinoids from the Middle Eocene Lake City Formation of Georgia. *Smithsonian Miscellaneous Collections*, **153**, 1–45.

Kikuchi, M., Omori, A., Kurokawa, D., and Akasaka, K. (2015). Patterning of anteroposterior body axis displayed in the expression of Hox genes in sea cucumber *Apostichopus japonicus*. *Development Genes and Evolution*, **225**, 275–286.

Kroh A., Lukeneder, A., and Gallemí, J. (2014). *Absurdaster*, a new genus of basal atelostomate from the Early Cretaceous of Europe and its phylogenetic position. *Cretaceous Research*, **48**, 235–249.

Lam, A. R., Sheffield, S. L., and Matzke, N. J., (2021). Estimating dispersal and evolutionary dynamics in diploporan blastozoans (Echinodermata) across the great Ordovician biodiversification event. *Paleobiology*, **47**, 198–220.

Lev, O., Edgecombe, G. D., and Chipman, A. D. (2022). Serial homology and segment identity in the arthropod head. *Integrative Organismal Biology*, **4**, obac015.

Linnaeus, C. (1758). *Systema naturae* (Vol. 1). Stockholm: Holmiae (Laurentii Salvii).

Lowe, C. J., and Wray, G. A. (1997). Radical alterations in the roles of homeobox genes during echinoderm evolution. *Nature*, **389**, 718–721.

McKinney, M. L., and Sumrall, C. D. (2011). Ambulacral growth allometry in edrioasteroids: Functional surface-volume change in ontogeny and phylogeny. *Lethaia*, **44**, 102–108.

Mooi, R., David, B., and Marchand, D. (1994). Echinoderm skeletal homologies: Classical morphology meets modern phylogenetics. In B. David, A. Guille, J. P. Féral, and M. Roux, eds., *Echinoderms through Time*. Balkema: Rotterdam, pp. 87–95.

Mooi, R. and David, B. (1997). Skeletal homologies of echinoderms. In J. A. Waters and G. G. Maples, eds. Geobiology of echinoderms. *Paleontological Society Papers*, **3**, 305–335.

Mooi, R., and David, B. (1998). Evolution within a bizarre phylum: Homologies of the first echinoderms. *American Zoologist*, **38**, 965–974.

Mooi, R., and David, B., 2000. What a new model of skeletal homologies tells us about asteroid evolution. *American Zoologist*, **40**, 326–339.

Mooi, R., David, B., and Wray, G. A. (2005). Arrays in rays: Terminal addition in echinoderms and its correlation with gene expression. *Evolution & Development*, **7**, 542–555.

Mooi, R., and David, B. (2008). Radial symmetry, the anterior/posterior axis, and echinoderm Hox genes. *Annual Review of Ecology, Evolution, & Systematics*, **39**, 43–62.

Morris, V. B. (2007). Origins of radial symmetry identified in an echinoderm during adult development and the inferred axes of ancestral bilateral symmetry. *Proceedings of the Royal Society B: Biological Sciences*, **274**, 1511–1516.

Morris, V. B. (2012). Early development of coelomic structures in an echinoderm larva and a similarity with coelomic structures in a chordate embryo. *Development genes and evolution*, **222**, 313–323.

Morris, V. B., and Byrne, M. (2005). Involvement of two Hox genes and Otx in echinoderm body-plan morphogenesis in the sea urchin *Holopneustes purpurescens*. *Journal of Experimental Zoology Part B: Molecular and Developmental Evolution*, **304**, 456–467.

Morris, V. B., and Byrne, M. (2014). Oral–aboral identity displayed in the expression of HpHox3 and HpHox11/13 in the adult rudiment of the sea urchin *Holopneustes purpurescens*. *Development Genes and Evolution*, **224**, 1–11.

Nohejlová, M., and Fatka, O. (2016). Ontogeny and morphology of Cambrian eocrinoid *Akadocrinus* (Barrandian area, Czech Republic). *Bulletin of Geosciences*, **91**, 141–153.

Okada, M. (1979). The central role of the genital duct in the development and regeneration of the genital organs in the sea urchin. *Development, Growth & Differentiation*, **21**, 567–576.

Oliveri, P., Tu, Q., and Davidson, E. H. (2008). Global regulatory logic for specification of an embryonic cell lineage. *Proceedings of the National Academy of Sciences*, **105**, 5955–5962.

O'Malley, C. E., Ausich, W. I., and Chin, Y. (2016). Deep echinoderm phylogeny preserved in organic molecules from Paleozoic fossils. *Geology*, **44**, 379–382.

Ortega-Hernández, J., Janssen, R., and Budd, G. E. (2017). Origin and evolution of the panarthropod head: A palaeobiological and developmental perspective. *Arthropod Structure & Development*, **46**, 354–379.

Nohejlová, M., Nardin E., Fatka, O., Kašička L., and Szabad M. (2019). Morphology, palaeoecology and phylogenetic interpretation of the Cambrian echinoderm *Vyscystis* (Barrandian area. Czech Republic). *Journal of Systematic Palaeontology* **17**, 1619–1634.

Patterson, C. (1982). Morphological characters and homology. In K. A. Joysey and A. E. Friday, eds., *Problems of Phylogenetic Reconstruction*. London and New York: Academic Press, pp. 21–74.

Parsley, R. L. (1970). Revision of the North American Pleurocystitidae (Rhombifera – Cystoidea). *Bulletins of American Paleontology*, **58**, 132–213.

Parsley, R. L. (1980). Homalozoa. In T. W. Broadhead and J. A. Waters, eds., *Echinoderms, Notes for a Short Course: Studies in Geology*, **3**, 106–117.

Parsley, R. L. (2013). Development and functional morphology of sutural pores in Early and Mid-Cambrian gogiid eocrinoids from Guizhou Province, China. In C. Johnson, ed., *Echinoderms in a Changing World: Proceedings of the 13th International Echinoderm Conference*. Hobart: University of Tasmania, pp. 79–86.

Parsley, R.L., and Mintz, L.W. (1975). North American Paracrinoidea: (Ordovician: Paracrinozoa, new, Echinodermata): Bulletins of American Paleontology, 68, 1–113.

Parsley, R. L., and Zhao, Y. (2006). Long stalked eocrinoids in the basal Middle Cambrian Kaili Biota, Taijiang County, Guizhou Province, China. *Journal of Paleontology*, **80**, 1058–1071.

Paul, C. R. C. (1967). The functional morphology and mode of life of the cystoid *Pleurocystites*, E. Billings, 1854. In N. Millott, ed., *Echinoderm Biology: Symposia of the Zoological Society of London*, **20**, 105–123.

Paul, C. R. C. (1968a). *Macrocystella* Callaway, the earliest glyptocystitid cystoid. *Palaeontology*, **11**, 580–600.

Paul, C. R. C. (1968b). Morphology and function of the dichoporite pore-structures in cystoids. *Palaeontology*, **11**, 697–730.

Paul, C. R. C. (1984). British Ordovician Cystoids Part 2. *Palaeontographical Society Monographs*, **563**, 65–152.

Paul, C. R. C. (2021). New insights into the origin and relationships of blastoid echinoderms. *Acta Palaeontologica Polonica*, **66**, 41–62. DOI: https://doi .org/10.4202/app.00825.2020.

Paul, C. R., and Hotchkiss, F. H. (2020). Origin and significance of Lovén's Law in echinoderms. *Journal of Paleontology*, **94**, 1089–1102.

Paul, C. R. C., and Toom, U. (2021). The diploporite blastozoan *Glyptosphaerites* (Echinodermata: Blastozoa) and the origin of diplopores. *Estonian Journal of Earth Sciences*, **70**, 224–239.

Peters, J., and Lane, N. G. (1990). Ontogenetic adaptations in some Pennsylvanian crinoids. *Journal of Paleontology*, **64**, 427–435.

Peterson, K. J., Arenas-Mena, C., and Davidson, E. H. (2000). The A/P axis in echinoderm ontogeny and evolution: evidence from fossils and molecules. *Evolution & development*, **2**, 93–101.

Piovani, L., Czarkwiani, A., Ferrario, C., Sugni, M., and Oliveri, P. (2021). Ultrastructural and molecular analysis of the origin and differentiation of cells mediating brittle star skeletal regeneration. *BMC Biology*, **19**, 1–19.

Rowe, T. (1988). Definition, diagnosis, and origin of Mammalia. *Journal of Vertebrate Paleontology*, **8**, 241–264.

Saucède, T., David, B., and Mooi, R. (2001). The strange apical system of the genus *Pourtalesia* (Holasteroida, Echinoidea). *Echinoderm Research*, 131–136.

Savriama, Y., Stige, L. C., Gerber, S. et al. (2015). Impact of sewage pollution on two species of sea urchins in the Mediterranean Sea (Cortiou, France): Radial asymmetry as a bioindicator of stress. *Ecological Indicators*, **54**, 39–47.

Shashikant, T., Khor, J. M., and Ettensohn, C. A. (2018). From genome to anatomy: The architecture and evolution of the skeletogenic gene regulatory network of sea urchins and other echinoderms. *Genesis*, **56**, e23253.

Sheffield, S. (2013). The Pennsylvanian cladid crinoid *Erisocrinus*: Ontogeny and systematics. Unpublished M.S. thesis, Auburn University.

Sheffield, S. L., and Sumrall, C. D. (2017). Generic revision of the Holocystitidae of North America (Diploporita: Echinodermata) based on universal elemental homology. *Journal of Paleontology*, **91**, 755–766. DOI: https://doi.org/10.1017/jpa.2016.159.

Sheffield, S. L., and Sumrall, C. D. (2019). The phylogeny of the Diploporita: A polyphyletic assemblage of blastozoan echinoderms. *Journal of Paleontology*, **93**, 740–752.

Sheffield, S. L., Sumrall, C. D., and Ausich, W. I. (2018). Late Ordovician (Hirnantian) diploporitan fauna of Anticosti Island, Quebec, Canada: Implications for evolutionary and biogeographic patterns. *Canadian Journal of Earth Sciences*, **55**, 1–7. DOI: https://doi.org/10.1139/cjes-2017-0160.

Sheffield, S. L., Limbeck, M. R., Bauer, J. E., Hill, S. A., and Nohejlová, M. (2022). A review of blastozoan echinoderm respiratory structures. Elements of Paleontology. Cambridge, UK: Cambridge University Press. doi:10.1017/9781108881821

Shubin, N. H., Tabin, C., and Carroll, S. (2009). Deep homology and the origins of evolutionary novelty. *Nature*, **457**, 818–823.

Shubin N. H., and Marshall C. R. (2000). Fossils, genes, and the origin of novelty. *Paleobiology*, **26**, 324–340.

Smith, A. B. (1984a). *Echinoid Palaeobiology*. Vol. 1. London: Allen & Unwin.

Smith, A. B. (1984b). Classification of the Echinodermata. *Palaeontology*, **27**, 431–459.

Smith, A. B. (1985). Cambrian eleutherozoan echinoderms and the early diversification of edrioasteroids. *Palaeontology*, **28**, 715–756.

Smith, A. B. (2004). Phylogeny and systematics of holasteroid echinoids and their migration into the deep-sea. *Palaeontology*, **47**, 123–150.

Smith, A. B. (2005). The pre-radial history of echinoderms. *Geological Journal*, **40**, 255–280.

Smith, A. B., and Jell, P. A. (1990). Cambrian edrioasteroids from Australia and the origin of starfishes. *Memoirs of the Queensland Museum*, **28**, 715–778.

Smith, A. B., and Zamora, S. (2013). Cambrian spiral-plated echinoderms from Gondwana reveal the earliest pentaradial body plan. *Proceedings of the Royal Society B: Biological Sciences*, **280**, 20131197.

Spirlet, C., Grosjean, P., and Jangoux, M. (1994). Differentiation of the genital apparatus in a juvenile echinoid (*Paracentrotus lividus*). In B. David, A. Guille, J. P. Féral, and M. Roux, eds., *Echinoderms through Time*. Balkema: Rotterdam, pp. 881–886.

Sprinkle, J. (1973). *Morphology and Evolution of Blastozoan Echinoderms*. Cambridge, MA: Harvard University Museum of Comparative Zoology Special Publication.

Sprinkle, J. (1975). The "arms" of *Caryocrinites*, a rhombiferan cystoid convergent on crinoids. *Journal of Paleontology*, **49**, 1062–1073.

Sprinkle, J., and Wahlman, G. P. (1994). New echinoderms from the Early Ordovician of west Texas. *Journal of Paleontology*, **68**, 324–338.

Sumrall, C. D. (1996). Late Paleozoic edrioasteroids (Echinodermata) from the North American midcontinent. *Journal of Paleontology*, **70**, 969–985.

Sumrall, C. D. (1997). The role of fossils in the phylogenetic reconstruction of Echinodermata. *The Paleontological Society Papers*, **3**, 267–288.

Sumrall, C. D. (2001). Paleoecology and taphonomy of two new edrioasteroids from a Mississippian hardground in Kentucky. *Journal of Paleontology*, **75**, 136–146.

Sumrall, C. D. (2010). A model for elemental homology for the peristome and ambulacra in blastozoan echinoderms. In L. G. Harris, S. A. Böttger, C. W. Walker, and M. P. Lesser, eds., *Echinoderms*. CRC Durham, London: CRC Press, pp. 269–276.

Sumrall, C. D. (2015). Understanding the oral area of derived stemmed echinoderms. In S. Zamora & I. Rábano, eds., *Progress in Echinoderm Palaeobiology: Cuademos del Museo Geominero. Madrid: Instituto Geológico y Minero de España*, **19**, pp.169–173.

Sumrall, C. D. (2017). New insights concerning homology of the oral region and ambulacral system plating of pentaradial echinoderms. *Journal of Paleontology*, **91**, 604–617.

Sumrall, C. D. (2020.) Echinodermata. In K. de Queiroz, J. Gauthier, and P. Cantino, eds., *Phylonyms: A Companion Volume to the PhyloCode*. London: Taylor & Francis Group, pp. 645–648.

Sumrall, C. D., Brett, C. E., Dexter, T. A., and Bartholomew, A. (2009). An enigmatic blastozoan echinoderm fauna from central Kentucky. *Journal of Paleontology*, **83**, 739–749.

Sumrall, C. D., and Gahn, F. J. (2006). Morphological and systematic reinterpretation of two enigmatic edrioasteroids (Echinodermata) from Canada. *Canadian Journal of Earth Sciences*, **43**, 497–507.

Sumrall, C. D., and Schumacher, G. A. (2002). *Cheirocystis fultonensis*, a new glyptocystitoid rhombiferan from the Upper Ordovician of the Cincinnati Arch – comments on cheirocrinid ontogeny. *Journal of Paleontology*, **76**, 843–851.

Sumrall, C. D., B. Deline, J. Colmenar, S. L. Sheffield, and S. Zamora. (2015). New data on late Ordovician (Katian) echinoderms from Sardinia, Italy. In S. Zamora and I. Rábano, eds., *Progress in Echinoderm Palaeobiology: Cuademos del Museo Geominero*. Madrid: Instituto Geológico y Minero de España, **19**, pp. 159–162.

Sumrall, C. D., and Phelps, D. (2021). *Spiracarneyella*, a new carneyellid edrioasteroid from the Upper Ordovician (Katian) of Kentucky and Ohio and comments on carneyellid heterochrony. *Journal of Paleontology*, **95**, 624–629.

Sumrall, C. D., and Sprinkle, J. (1995). Plating and pectinirhombs of the Ordovician rhombiferan *Plethoschisma*. *Journal of Paleontology*, **69**, 772–778.

Sumrall, C. D., and Sprinkle, J. (1999). Early ontogeny of the glyptocystitid rhombiferan *Lepadocystis moorei*. In M. D. C. Carnevali and F. Bonasoro, eds., *Echinoderm Research 1998*. Rotterdam: Balkema, pp. 409–414.

Sumrall, C. D., Sprinkle, J., and Guensburg, T. E. (2001). Comparison of flattened blastozoan echinoderms: Insights from the new Early Ordovician eocrinoid *Haimacystis rozhnovi*. *Journal of Paleontology*, **75**, 985–992.

Sumrall, C. D., and Waters, J. A. (2012). Universal elemental homology in glyptocystitoids, hemicosmitoids, coronoids and blastoids: Steps toward echinoderm phylogenetic reconstruction in derived Blastozoa. *Journal of Paleontology*, **86**, 956–972.

Sumrall, C. D., and Wray, G. A. (2007). Ontogeny in the fossil record: Diversification of body plans and the evolution of "aberrant" symmetry in Paleozoic echinoderms. *Paleobiology*, **33**, 149–163.

Sumrall, C. D., and Zamora, S. (2011). Ordovician edrioasteroids from Morocco: Faunal exchanges across the Rheic Ocean. *Journal of Systematic Palaeontology*, **9**, 425–454.

Sumrall, C. D., and Zamora, S. (2018). New Upper Ordovician edrioasteroids from Morocco. *Geological Society, London, Special Publications*, **485**, 565–577.

Thompson, J. (2022). Molecular Paleobiology of the Echinoderm Skeleton Elements of Paleontology. Cambridge, UK: Cambridge University Press. doi:10.1017/9781009179768

Thompson, J. R., Cotton, L. J., Candela, Y. et al. (2022). The Ordovician diversification of sea urchins: Systematics of the Bothriocidaroida (Echinodermata: Echinoidea). Journal of Systematic Palaeontology, **19**, 1395–1448.

Thompson, J. R., Paganos, P., Benvenuto, G., Arnone, M. I., and Oliveri, P. (2021). Post-metamorphic skeletal growth in the sea urchin *Paracentrotus lividus* and implications for body plan evolution. *EvoDevo*, **12**, 1–14.

Tsuchimoto, J., and Yamaguchi, M. (2014). Hox expression in the direct-type developing sand dollar *Peronella japonica*. *Developmental Dynamics*, **243**, 1020–1029.

Tweedt, S. M. (2017). Gene regulatory networks, homology, and the early panarthropod fossil record. *Integrative and Comparative Biology*, **57**, 477–487.

Ubaghs, G. (1971). Diversité et spécialisation des plus anciens Échinodermes que l'on connaisse. *Biological Reviews*, **46**, 157–200.

Wagner, G. P. (2007) The developmental genetics of homology. *Nature Reviews Genetics*, **8**, 473–479.

Wright, D. F. (2015). Fossils, homology, and "Phylogenetic Paleo-ontogeny": A reassessment of primary posterior plate homologies among fossil and living crinoids with insights from developmental biology. *Paleobiology*, **41**, 570–591.

Zamora, S., Linán, E., Alonso, P. D., Gozalo, R., and Vintaned, J. A. G. (2007). A Middle Cambrian edrioasteroid from the Murero biota (NE Spain) with Australian affinities. *Annales de Paléontologie*, **93**, 249–260).

Zamora, S., and Rahman, I. A. (2014). Deciphering the early evolution of echinoderms with Cambrian fossils. *Palaeontology*, **57**, 1105–1119.

Zamora, S., Rahman, I. A., and Smith, A. B. (2012). Plated Cambrian bilaterians reveal the earliest stages of echinoderm evolution. *PLoS One*, **7**, e38296.

Zamora, S., and Sumrall, C.D. In press. Morphology and relationships of early pentaradial echinoderms. Elements of Paleontology. Cambridge, UK: Cambridge University Press.

Zamora, S., Rahman, I. A., Sumrall, C. D., Gibson, A. P., and Thompson, J. R. (2022). Cambrian edrioasteroid reveals new mechanism for secondary reduction of the skeleton in echinoderms. *Proceedings of the Royal Society B*, **289**, 20212733.

Zamora, S., Sumrall, C. D., and Vizcaïno, D. (2012). Morphology and ontogeny of the Cambrian edrioasteroid echinoderm *Cambraster cannati* from western Gondwana. *Acta Palaeontologica Polonica*, **58**, 545–559.

Zamora, S., Sumrall, C. D., Zhu, X. -J., and Lefebvre, B. (2017). A new stemmed echinoderm from the Furongian of China and the origin of Glyptocystitida (Blastozoa, Echinodermata). *Geological Magazine*, **154**, 465–475.

Zhao, Y., Sumrall, C. D., Parsley, R. L., and Peng, J. (2010). *Kailidiscus*, a new plesiomorphic edrioasteroid from the basal Middle Cambrian Kaili biota of Guizhou Province, China. *Journal of Paleontology*, **84**, 668–680.

Acknowledgments

We thank reviewers E. Nardin and D. F. Wright for their comments that improved this Element, as well as editorial comments from S. Zamora. We also thank M. Limbeck, S. Hill, and W. Lapic for early feedback. For access to specimens, we thank the collections staff at the museums listed in the institutional abbreviations. We thank M. Nohejlová for providing images. JRT was funded by a Leverhulme Trust Early Career Fellowship.

Cambridge Elements

Elements of Paleontology

About the Series

The Elements of Paleontology series is a publishing collaboration between the Paleontological Society and Cambridge University Press. The series covers the full spectrum of topics in paleontology and paleobiology, and related topics in the Earth and life sciences of interest to students and researchers of paleontology.

The Paleontological Society is an international nonprofit organization devoted exclusively to the science of paleontology: invertebrate and vertebrate paleontology, micropaleontology, and paleobotany. The Society's mission is to advance the study of the fossil record through scientific research, education, and advocacy. Its vision is to be a leading global advocate for understanding life's history and evolution. The Society has several membership categories, including regular, amateur/avocational, student, and retired. Members, representing some forty countries, include professional paleontologists, academicians, science editors, Earth science teachers, museum specialists, undergraduate and graduate students, postdoctoral scholars, and amateur/avocational paleontologists.

Cambridge Elements

Elements of Paleontology

Elements in the Series

Functional Micromorphology of the Echinoderm Skeleton
Przemyslaw Gorzelak

Disarticulation and Preservation of Fossil Echinoderms: Recognition of EcologicalTime Information in the Echinoderm Fossil Record
William I. Ausich

Crinoid Feeding Strategies: New Insights From Subsea Video And Time-Lapse
David Meyer, Margaret Veitch, Charles G. Messing and Angela Stevenson

Phylogenetic Comparative Methods: A User's Guide for Paleontologists
Laura C. Soul and David F. Wright

Expanded Sampling Across Ontogeny in Deltasuchus motherali *(Neosuchia, Crocodyliformes): Revealing Ecomorphological Niche Partitioning and Appalachian Endemism in Cenomanian Crocodyliforms*
Stephanie K. Drumheller, Thomas L. Adams, Hannah Maddox and Christopher R. Noto

Testing Character Evolution Models in Phylogenetic Paleobiology: A Case Study with Cambrian Echinoderms
April Wright, Peter J. Wagner and David F. Wright

The Taphonomy of Echinoids: Skeletal Morphologies, Environmental Factors and Preservation Pathways
James H. Nebelsick and Andrea Mancosu

Virtual Paleontology: Tomographic Techniques For Studying Fossil Echinoderms
Jennifer E. Bauer and Imran A. Rahman

Follow The Fossils: Developing Metrics For Instagram As A Natural Science Communication Tool
Samantha B. Ocon, Lisa Lundgren, Richard T. Bex II, Jennifer E. Bauer, Mary Janes Hughes, and Sadie M. Mills

Niche Evolution and Phylogenetic Community Paleoecology of Late Ordovician Crinoids
Selina R. Cole and David F. Wright

Molecular Paleobiology of the Echinoderm Skeleton
Jeffrey R. Thompson

A Review of Blastozoan Echinoderm Respiratory Structures
Sarah L. Sheffield, Maggie R. Limbeck, Jennifer E. Bauer, Stephen A. Hill and Martina Nohejlová

A full series listing is available at: www.cambridge.org/EPLY

Printed in the United States
by Baker & Taylor Publisher Services